Understanding Voltammetry:
Simulation of Electrode Processes

Second Edition

Other World Scientific Titles by the Author

Understanding Voltammetry
(Third Edition)
by Richard G Compton and Craig E Banks
ISBN: 978-1-78634-526-4
ISBN: 978-1-78634-529-5 (pbk)

Understanding Voltammetry: Problems and Solutions
by Richard G Compton, Christopher Batchelor-McAuley and
Edmund J F Dickinson
ISBN: 978-1-84816-730-8
ISBN: 978-1-84816-731-5 (pbk)

A. G. Stromberg — First Class Scientist, Second Class Citizen:
Letters from the GULAG and a History of Electroanalysis in the USSR
by Richard G Compton, Alexander S Kabakaev, Michael T Stawpert,
Gregory G Wildgoose and Elza A Zakharova
ISBN: 978-1-84816-675-2

Understanding Voltammetry:
Simulation of Electrode Processes

Second Edition

Richard G Compton
Enno Kätelhön
Kristopher R Ward
Oxford University, UK

Eduardo Laborda
University of Murcia, Spain

 World Scientific

NEW JERSEY · LONDON · SINGAPORE · BEIJING · SHANGHAI · HONG KONG · TAIPEI · CHENNAI · TOKYO

Published by

World Scientific Publishing Europe Ltd.

57 Shelton Street, Covent Garden, London WC2H 9HE

Head office: 5 Toh Tuck Link, Singapore 596224

USA office: 27 Warren Street, Suite 401-402, Hackensack, NJ 07601

Library of Congress Cataloging-in-Publication Data

Names: Compton, R. G., author. | Kätelhön, Enno, author. |
    Ward, Kristopher R., author. | Laborda, Eduardo, 1983–   author.
Title: Understanding voltammetry : simulation of electrode processes /
    Richard G. Compton, Oxford University, Enno Kätelhön, Oxford University,
    Kristopher R. Ward, Oxford University, UK, Eduardo Laborda, University of Murcia, Spain.
Description: Second edition. | New Jersey : World Scientific, [2020] |
    Includes bibliographical references and index.
Identifiers: LCCN 2019052167 | ISBN 9781786348302 (hardcover)
Subjects: LCSH: Voltammetry--Textbooks. | Electrochemistry--Textbooks.
Classification: LCC QD116.V64 C66 2020 | DDC 543/.4--dc23
LC record available at https://lccn.loc.gov/2019052167

British Library Cataloguing-in-Publication Data
A catalogue record for this book is available from the British Library.

For any available supplementary material, please visit
https://www.worldscientific.com/worldscibooks/10.1142/Q0246#t=suppl

Printed in Singapore

Preface to the Second Edition

Since the publication of the 1st Edition of *Understanding Voltammetry: Simulation of Electrode Processes* in 2013 we have extensively used our book for teaching, especially new research students, and also received much constructive feedback from our readers. All this has guided the enlargement and revision of the text culminating in the 2nd Edition. In particular we have added significant material on random walks and have provided more material on the checking and validation of simulations. Pulse techniques, and particularly square wave voltammetry, receive greater attention than previously. Moreover the text has been revised throughout to hopefully provide greater clarity and insight especially at points of potential difficulty.

We note that the 3rd Edition of the companion text *Understanding Voltammetry* (World Scientific) was published in 2018.

RGC, EK, EL, KRW, October 2019

Preface to the First Edition

This book is aimed at students, researchers and professors in the broad area of electrochemistry who wish to simulate electrochemical processes in general, and voltammetry in particular. It has been written as a result of numerous requests over recent years for assistance in getting started on such activity and aims to lead the novice reader who has some prior experience of experimental electrochemistry at least, from a state of zero knowledge to being realistically able to embark on using simulation to explore non-standard experiments using cyclic voltammetry, microdisc electrodes and hydrodynamic electrodes such as rotating discs.

The book complements the texts *Understanding Voltammetry* (2nd edition, Imperial College Press, 2011, written with C. E. Banks) and *Understanding Voltammetry: Problems and Solutions* (Imperial College Press, 2012, with C. Batchelor-McAuley and E. J. F. Dickinson).

We wish the reader many happy hours of simulation and hope that this is fully repaid by the impact on their experimental research.

RGC, EL, KRW, July 2013

Contents

Chapter 1

Introduction

In any scientific field obtaining information about an experimental system requires the appropriate model of its response. Modelling also enables us to understand and predict the experimental behaviour in order to ensure optimum experimental conditions in terms of sensitivity and minimisation of undesirable effects.

Generally speaking the mathematical problems tackled in voltammetry involve the resolution of partial differential equation systems by means of analytical, semi-analytical or numerical methods. The solutions of the problem are the concentration profiles of the different species, and from them the current-potential-time response of the system to a given electrical perturbation can be calculated.

Analytical methods provide exact solutions that allow for direct analysis of the influence of experimental variables and the determination of the conditions for particular behaviours such as the achievement of a steady-state signal. Nevertheless the use of analytical methods is not always feasible due to the complexity of the problems. In such cases numerical methods offer a very accurate approximation to the true solution once the conditions of the simulation are optimised.

Unfortunately, simulation is usually obscure for non-theoreticians who often have to rely on "black-box" software packages. This book aims to introduce the simulation of electrochemical experiments by numerical methods in a way that allows any reader to develop their own research and academic tools for the study of voltammetry. Rather than an exhaustive compilation of the existing modelling techniques, this is an introductory guide to electrochemical simulation where not difficult but accurate numerical methods are described for the most common situations in electrochemical studies and alternatives are mentioned where appropriate. Moreover, the reader can find specific advice on the computational implementation of the numerical techniques, including some coding examples with C++ used as the programming language. C++ is a highly flexible general-purpose computer programming language that has been used by a very large community in numerous disciplines over many years. Consequently, it is mature and well understood, and there is an abundance of resources such as tutorials and source code available to assist the novice. The code examples are intended to be easy to understand and so a lack of knowledge of the specifics of C++ should not deter the reader; the code should be simple to translate into any other suitable programming language. A brief review of some of the key features of the language is included in Appendix A.

While C++ may have a slightly steeper learning curve than some other languages traditionally used for numerical simulations, its flexibility allows for the development of not just single-purpose simulations but more powerful general-purpose simulation packages through the use of object-oriented programming; however, such methods are beyond the scope of this text.

The modelling of voltammetric experiments requires the definition of the system under study (in terms of mass transport, boundary conditions and heterogeneous/homogeneous chemical reactions) as well as of the electrical perturbation applied. These factors will obviously define the electrochemical response but also the optimum numerical method to employ. In the following chapters, general procedures for the easy implementation of methods to solve different electrochemical problems will be given along with indications for their optimisation and validation.

1.1. Electrochemical Systems

Electrochemists study chemical reactions which involve electron exchange processes for a variety of purposes, including chemical and biochemical sensing (e.g., glucose sensors, gas detectors, pH meters), technological applications (e.g., electroplating, electrochromic displays), energy storage (e.g., solar cells, batteries), imaging, synthesis, which underpin much of modern biology and nanotechnology.

These reactions can involve species in the same phase (homogeneous electron transfer reactions) or the electron can transfer through an interface (heterogeneous electron transfer reactions). In the second case, the transfer can occur between molecules (e.g., electron transfers at liquid-liquid interfaces or mediated by redox active monolayers) or between an electronic conductor (the electrode) and a molecule.

The last case is of great scientific interest given that we can act directly and readily on the energy of the electronic levels of the electrode through an external power supply (potentiostat/galvanostat). In this way, we can act on the kinetics and thermodynamics of the process. Let us consider the generic case of a one-electron electrode reaction according to the following scheme:

$$A^{z_A} + e^- \underset{k_b}{\overset{k_f}{\rightleftharpoons}} B^{z_A-1} \quad (E_f^0) \tag{1.1}$$

where z_A is the charge on species A, k_f and k_b the reduction and oxidation rate constants, respectively, and E_f^0 the formal potential of the redox couple A/B (see Eq. (1.8)).

We can "visualise" the electron transfer process (1.1) through the Gerischer-inspired diagrams [1] shown in Figure 1.1. The available electronic levels at the metal electrode are described according to its band structure with the Fermi level (E_F) defining the limit between the occupied (shaded area) and empty levels.[1] The curves on the right-hand side of the diagrams correspond to the distribution functions of electronic states available on the electroactive species at the electron transfer site (i.e., near the electrode surface), where the curve of the oxidised species A relates to empty states and that of the reduced species B to occupied levels. In Figure 1.1 equal concentrations for both species are assumed such that according to the Nernst equation (Eq. (1.9)) the equilibrium potential corresponds to the formal potential of the redox couple and the amplitudes of the density of occupied and empty states in solution are the same.

Electron transfer can take place from an occupied level of one of the phases (metal or solution) to an empty level of the other phase with the same energy. So, the reduction of species A into B corresponds to the electron transfer from an occupied state on the electrode to an empty state of species A in solution. The transfer from an occupied level of species B to empty states on the electrode represents the oxidation of species B.

The rate of the reduction and oxidation processes is related to the magnitude of the overlap of distribution functions of appropriate electronic states for the corresponding reaction. As can be seen in Figure 1.1(a), when the potential applied at the working electrode corresponds to the equilibrium potential, the overlap of occupied states on the electrode with (empty) states of species A is the same as that of empty states on the electrode with occupied states of species B. This reflects the fact that the rates of oxidation and reduction are the same and so a dynamic equilibrium is established.

---

[1] For the sake of simplicity we will consider absolute zero temperature such that the highest filled state corresponds to the Fermi level.

Fig. 1.1. Schematic of the distribution function of occupied (shaded) and empty (shadeless) electronic states at the interface between a metal electrode and a redox couple A/B in solution at equal concentrations of the electroactive species. The applied potential corresponds to (a) equilibrium, (b) cathodic (electroreduction) and (c) anodic (electrooxidation) conditions.

When the applied potential is changed to more negative values (Figure 1.1(b)), the electronic states at the electrode move towards higher energy, enhancing the overlap between occupied electrode states and states on species A. Consequently, the reduction of species A is promoted versus the oxidation of B. The opposite situation is obtained when applying a potential more positive than the equilibrium one (Figure 1.1(c)), which enhances the electron transfer from species B to electrode states.

As a result of the electrode reaction, changes in the composition of the electrolytic solution containing A and B will take place, and this will result in gradients of the electrochemical potential of the electroactive species $j \equiv$ A, B:

$$\bar{\mu}_j = \mu_j^0 + \mathcal{R}\mathcal{T}\ln\left(\frac{\gamma_j\,[j]}{[\,]^0}\right) + z_j\text{F}\phi_\text{S} \tag{1.2}$$

where μ_j^0 is the standard chemical potential of the species j, $\mathcal{R}$ the universal gas constant (8.314 J K^{-1} mol^{-1} in SI), $\mathcal{T}$ the absolute temperature (K), γ_j the activity coefficient of species j, $[j]$ its concentration (mol m^{-3}), $[\,]^0$ a standard concentration taken to be one molar, z_j the charge on the molecule, F the Faraday constant (96485 C mol^{-1}) and ϕ_S the electric potential in solution (V).

The electrochemical gradient is the driving force[2] for the response of the system to re-establish the electrochemical equilibrium that will involve the interplay of mass transport, (electro)chemical reactions and external forces.

The value of the equilibrium potential (E_eq) shown in Figure 1.1 can be derived from Eq. (1.2). Thus, electrochemical equilibrium conditions for the process (1.1) at constant temperature and pressure imply that

$$\bar{\mu}_\text{A} + \bar{\mu}_{\text{e}^-} = \bar{\mu}_\text{B} \tag{1.3}$$

which, according to the definition of the electrochemical potential (Eq. (1.2)), turns into

$$\left(\mu_\text{A}^0 + \mathcal{R}\mathcal{T}\ln\left(\frac{\gamma_\text{A}\,[\text{A}]}{[\,]^0}\right) + z_\text{A}\text{F}\phi_\text{S}\right) + \left(\mu_{\text{e}^-}^0 - \text{F}\phi_\text{M}\right)$$

$$= \mu_\text{B}^0 + \mathcal{R}\mathcal{T}\ln\left(\frac{\gamma_\text{B}\,[\text{B}]}{[\,]^0}\right) + (z_\text{A} - 1)\,\text{F}\phi_\text{S} \tag{1.4}$$

---

[2] For long duration experiments, $t > 10{-}20$ s, natural convection due to gradients of density can also take place.

where ϕ_S and ϕ_M refer to the electric potential at the solution and the electrode, respectively. From Eq. (1.4) it can be deduced that, under electrochemical equilibrium conditions, the interfacial potential difference, $\Delta\phi = \phi_M - \phi_S$, is given by the Nernst equation for a single electrode/solution interface:

$$\Delta\phi = \phi_M - \phi_S = \frac{\Delta\mu^0}{F} + \frac{RT}{F}\ln\left(\frac{\gamma_A [A]}{\gamma_B [B]}\right) \tag{1.5}$$

where

$$\Delta\mu^0 = \mu_A^0 + \mu_{e-}^0 - \mu_B^0 \tag{1.6}$$

The absolute value of $\Delta\phi$ is not accessible experimentally since for its measurement other interfaces are necessarily introduced. Instead its value can be measured relative to another electrode/solution interface with a constant potential difference ($\Delta\phi_{ref}$) that is provided by the reference electrode:

$$\Delta\phi - \Delta\phi_{ref} = \frac{\Delta\mu^0}{F} - \Delta\phi_{ref} + \frac{RT}{F}\ln\left(\frac{\gamma_A [A]}{\gamma_B [B]}\right) \tag{1.7}$$

The term on the left-hand side in Eq. (1.7) is the potential of the electrochemical equilibrium A/B at the working electrode with respect to the reference electrode. The first two terms on the right-hand side correspond to the equilibrium potential (versus the reference electrode potential) when the activities of the electroactive species are unity: $\gamma_j [j] = 1$. When measured relative to a standard hydrogen electrode (SHE) this is the so-called standard electrode potential, E^0, the value of which is characteristic of the redox couple A/B for given temperature.

Typically the activity coefficients of the electroactive species are unknown and therefore it is more practical to work with concentrations. With this aim, the concept of the formal potential, E_f^0, is introduced

$$E_f^0 = E^0 + \frac{RT}{F}\ln\left(\frac{\gamma_A}{\gamma_B}\right) \tag{1.8}$$

such that finally the well-known form of the Nernst equation is obtained:

$$E = E_f^0 + \frac{RT}{F}\ln\left(\frac{[A]}{[B]}\right) \tag{1.9}$$

Note that the value of the formal potential depends not only on the nature of the redox couple but also on the ionic strength of the electrolytic solution. Therefore, the E_f^0 value will be different in different media.

1.1.1. *Mass transport*

The active mechanisms of mass transport depend on the characteristics of the electrochemical experiments. In typical conditions, the cell consists of a static working electrode immersed in a stagnant solution containing the electroactive species as well as an excess of supporting electrolyte (an inert salt). The electrochemical reaction under study takes place at the surface of the working electrode, the potential of which is monitored with respect to the reference electrode. Finally, the counter (or auxiliary) electrode completes an electrical circuit with the working electrode over which the current flows. The use of this third electrode seeks to prevent large currents from passing through the reference electrode and causing variations of its potential. The electrochemical signal recorded reflects the process taking place at the working electrode such that our interest is focused on the interface of this electrode with the electrolytic solution.

The large quantity of supporting electrolyte added restrains the potential gradient to a region of a few ångström thickness from the electrode surface. Accordingly, the third term of Eq. (1.2), which is associated with the electric energy of the species, will be the same at any point of the electrolytic solution. On the other hand, there will exist a chemical gradient as a consequence of the different concentrations of the electroactive species depending on the distance to the electrode surface. The system will respond in order to balance the concentration gradient with the corresponding flux of material by diffusion.

For typical electrochemical experiments, where a very large number of molecules are involved, the diffusion process can be described by the statistical Fick's laws [2–4] which account for the changes in concentration with time and location.[3] Depending on the geometry of the electroactive surface (i.e., the electrode or array of electrodes acting as working electrode), the symmetry of the diffusion field may enable us to simplify the tridimensional problem to one or two dimensions.

The first and simplest case corresponds to linear, spherical and cylindrical diffusion associated with the use of planar, (hemi)spherical and cylindrical electrodes, respectively. As can be inferred from Figure 1.2, in these diffusion fields all the points at a given distance from the electrode surface in the perpendicular coordinate (x or r) are equivalent (*uniformly accessible*

---

[3] Within recent years more attention has been addressed to single-molecule electrochemistry. The simulation of these systems with a very low number of electroactive molecules requires stochastic modelling (see Chapter 11) [5].

Planar electrode   Hemispherical electrode   Cylindrical electrode

Fig. 1.2. Symmetry of the diffusion field at planar, (hemi)spherical and cylindrical electrodes.

electrodes) and net flux of molecules only occurs in this direction. According to Fick's first law, the flux, j $(\mathrm{mol\,m^{-2}\,s^{-1}})$, at a given point x_1, is proportional to the concentration gradient:

$$j_{j,x_1} = -D_j \left(\frac{\partial c_j}{\partial x} \right)_{x_1} \qquad (1.10)$$

where D_j $(\mathrm{m^2\,s^{-1}})$ is the diffusion coefficient of species j. Therefore, the problem simplifies and the solutions (i.e., the concentration profiles of the participating species) are functions of only two independent variables: time (t) and the distance to the electrode surface in the x- or r-coordinate.

From Eq. (1.10), Fick's second law can be derived which describes the change in concentration of species j with time. For the three cases considered in Figure 1.2 this law is given by

planar electrodes,

$$\frac{\partial c_j}{\partial t} = D_j \left(\frac{\partial^2 c_j}{\partial x^2} \right) \qquad (1.11)$$

(hemi)spherical electrodes,

$$\frac{\partial c_j}{\partial t} = D_j \left(\frac{\partial^2 c_j}{\partial r^2} + \frac{2}{r} \frac{\partial c_j}{\partial r} \right) \qquad (1.12)$$

cylindrical electrodes,

$$\frac{\partial c_j}{\partial t} = D_j \left(\frac{\partial^2 c_j}{\partial r^2} + \frac{1}{r} \frac{\partial c_j}{\partial r} \right) \qquad (1.13)$$

Let us consider an electrochemical experiment where after a time $t \geq 0$, the electroactive species A is reduced to B so that $c_A(x = 0, t) = 0$. If c_A is initially uniform and equal to its bulk value, this is the well-known Cottrell experiment [6]. Figure 1.3 shows the evolution with time of the concentration profiles of species A and B in this experiment under semi-infinite linear diffusion. Due to the decrease in concentration of species A near the electrode surface, molecules of A will flow from the bulk solution towards the electrode. On the other hand the species B electrogenerated at the electrode surface will diffuse towards the bulk solution. The region of solution where concentration changes happen (the so-called *diffusion layer*) will grow as time progresses. The growth rate depends on the diffusivity of the species in the medium, which is parameterised by its diffusion coefficient, D_j. The greater the diffusion coefficient is, the faster the spreading of the disturbed region.

The flux of species A at the electrode-solution interface $(x = 0)$ is particularly important in electrochemical measurements given that it is directly related to the rate of transformation of species A and therefore to the current that flows through the working electrode, I:[4]

$$\frac{I}{F A} = -D_A \left(\frac{\partial c_A}{\partial x} \right)_{x=0} \qquad (1.14)$$

where A is the area of the working electrode. As can be seen in Figure 1.3, the concentration gradient and therefore the surface flux of each species decrease with time as the profiles are less steep.

An estimation of the thickness of the diffusion layer can be obtained through the Nernst diffusion layer, δ (m), defined as shown in Figure 1.3. For the Cottrell experiment under linear diffusion conditions the δ value for a species j is given by

$$\delta_j = \sqrt{\pi D_j t} \qquad (1.15)$$

where t (s) is the time of the experiment. This expression is very important since it gives us an idea of the extent of the region of solution where the concentration changes take place. Thus, typically the region of space that is considered when modelling an electrochemical experiment extends up to a distance $6\sqrt{D_{max} t_{max}}$ (where D_{max} is the greatest diffusion coefficient of

---

[4] Following the IUPAC recommendations for sign conventions of electrochemical data, positive values will be assigned to anodic currents (electrooxidation) and negative ones to cathodic currents (electroreduction).

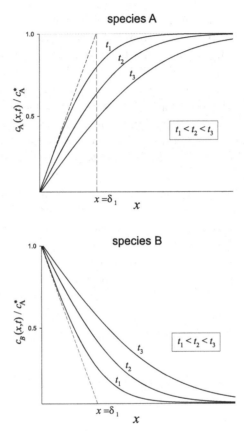

Fig. 1.3. Variation with time of the concentration profiles of species A and B in the Cottrell experiment (see text) at a planar macroelectrode. The Nernst diffusion layer for t_1 (δ_1) is indicated on the graph. c_A^* is the bulk concentration of species A. $D_A = D_B$.

the species involved in the problem and t_{max} is the total duration of the experiment) from the electrode surface to guarantee that all the region with concentration gradients is covered.

Unfortunately the use of planar and (hemi)spherical electrodes is not always appropriate or possible in electrochemical studies. Electrodes with large areas lead to problems derived from large ohmic drop and capacitive effects, and the fabrication of (hemi)spherical microelectrodes is difficult. Consequently micro*disc* electrodes are ubiquitous in electrochemical experiments since they allow for the reduction of the above undesirable effects and are easy to manufacture and clean. This is also true in the case of band electrodes and electrodes with heterogeneous surfaces due to the non-uniform nature of the electrode material (e.g., basal plane pyrolytic graphite electrodes) and in the case of the modification of the surface with nanoparticles, porous films, etc. The modelling of electrochemical experiments with these electrodes obliges us to deal with two-dimensional problems (Figure 1.4) and non-uniform surface fluxes which complicate the numerical resolution of the problem. Procedures for the simulation of two-dimensional electrochemical problems will be given in Chapters 9 and 10.

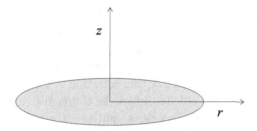

Fig. 1.4. Coordinate system for a microdisc electrode.

Advances in electrochemistry have led to deviations from the classical setup of electrochemical experiments [5]. The absence of an excess of supporting electrolyte, that is, electrochemical experiments in weakly supported media, is one of them. An excess of supporting electrolyte is usually added to the electrolytic solution to contain the extent of the electrical double layer and to compensate for the charge generated upon the electrode reaction. However, this is not always possible or desirable. The addition of an inert salt can introduce impurities and affect the behaviour of the electroactive species through the ionic strength (e.g., biomolecules) or chemical reactions (e.g., complexation, ion-pairing). Moreover the supporting electrolyte may not be soluble in the amount required, which is common in non-polar solvents. In stripping experiments, fully supported conditions may not be achieved due to the high local concentrations near the electrode of the ionic species generated during the release step.[5] Finally and most

---

[5] Stripping voltammetries are very sensitive electroanalytical methods where the analyte is preconcentrated in the electrode (mercury electrodes) or on its surface (solid electrodes) and subsequently released by applying a potential-time program as those described in Section 1.2.

importantly, extra kinetic and mechanistic information is often available if voltammetry is conducted with variable levels of supporting electrolyte [7]. For all the above reasons it is necessary to consider the situation where a local potential gradient (electric field) occurs in the vicinity of the electrode surface. Consequently, the potential in solution at the site of the electron transfer is different from its value in bulk solution. This affects the mass transport of the electroactive species and the real driving force experienced by the reactant (resistive or ohmic drop effects).

Regarding the mass transport, electrically charged species will migrate towards the electrode surface or the bulk solution in order to disperse the charge created by the electrode process. In a system with a large concentration of supporting electrolyte,[6] the migration of the added ionic species cancels the potential gradient around the electrode surface such that the transport by migration of the electroactive species can be considered negligible. On the other hand, in a low concentration of supporting electrolyte, the added species are less able to distort the charge distribution and so the migration of the electroactive species to mitigate the potential drop will be significant.

The migrational contribution to the flux of species may be in the same or opposite direction as the diffusional flux, depending on the sign of the charge of the electrode surface, the excess of charge generated around the electrode surface, and the charge of the electroactive species. In Figure 1.5, the reduction of species A is considered when applying a potential more negative than the potential of zero charge (E_{PZC}) such that the electrode surface is negatively charged. As can be seen in Figure 1.5(a), in the reduction of cations, migration contributes to the flux towards the electrode surface of reactant A, which is attracted by the negatively charged electrode surface and the defect of positive charge generated upon the electron transfer reaction. The opposite situation is observed for the reduction of anions (Figure 1.5(b)). Consequently, the reductive current (proportional to the surface flux of species A) with respect to purely diffusional conditions will be higher in the reduction of cations and smaller in the reduction of anions. This will be discussed in detail in Chapter 7.

Apart from diffusion and migration, transport by convection can also take place due to different internal and external forces. Thus, *natural con-*

---

[6] The concentration of the supporting electrolyte must be more than 100 times that of the electroactive species to ensure purely diffusional conditions in cyclic voltammetry at macroelectrodes [8].

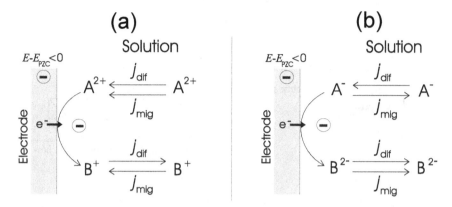

Fig. 1.5. Contributions of diffusion (j_{dif}) and migration (j_{mig}) to the mass transport of electroactive species for the one-electron reduction of the dication A^{2+} (a) and the monoanion A^- (b). The applied potential is supposed to be more negative than the potential of zero charge: $E - E_{\text{PZC}} < 0$.

vection due to gradients of density can occur when the electrode reaction provokes a significant local change in the solution composition or due to thermal variations. The modelling of this case is difficult[7] and so electrochemical experiments are usually restricted to short time scales, low concentration of analyte, and thermostated cells such that the influence of natural convection is minimised.

Forced convection as a result of the application of external forces to the system is more useful and simpler to model. The aim is the enhancement of mass transport of electroactive species towards the electrode surface in order to have better sensitivity and determine faster rate constants. Moreover, steady-state conditions are achieved quickly such that more reproducible results are obtained and charging current effects are minimised. Examples of controlled convective systems include the use of non-static electrodes (e.g., rotating disc and dropping mercury electrodes) and non-static electrolytic solutions (e.g., channel, tubular, wall-jet and wall-tube electrodes for analysis of flowing solutions). Given that the relative movement of the solution with respect to the electrode is well controlled and known, it can be modelled accurately by modifying the differential equations with the corresponding convective terms. The cases of the rotating disc and channel electrodes (Figure 1.6) will be studied in Chapter 8.

---

[7] However, in the last decade significant progress has been made for the appropriate description of natural convection [9, 10].

a)

Fig. 1.6. Schemes of the flow of solution in the (a) rotating disc and (b) channel electrodes.

1.1.2. *Boundary conditions*

For the resolution of the partial differential equations defining mass transport there are a number of conditions that the solutions must meet. These will be examined in detail for each particular problem throughout the book but some general guidelines are given in this section (Figure 1.7).

The number of these conditions for each dependent variable (the concentration profile of each species) depends on the number of independent variables and the order of the differential equation (i.e., the highest derivative of the dependent variable with respect to the independent one).

The partial differential equation corresponding to Fick's second law is first-order in time such that one temporal boundary condition is required for each species. This is the initial condition of the system, that is, the concentration profiles at the beginning of the electrochemical experiment, $t = 0$. For example, if the system is at equilibrium then the concentration of any species is uniform across the solution such that the initial condition is

$$c_j(x, t = 0) = c_j^* \tag{1.16}$$

where c_j^* is the initial uniform bulk concentration of species j in solution.

With respect to the spatial derivatives, taking for example the case of a planar electrode (Eq. (1.11)), two boundary conditions are needed, one at each end of the concentration profiles. Thus, a limit condition is associated with the point of the simulation space farthest away from the

Solution

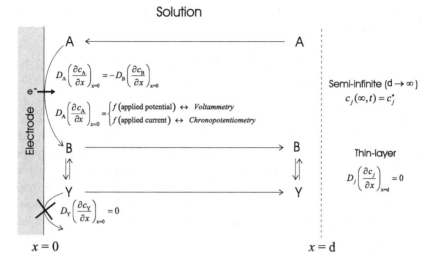

Fig. 1.7. Scheme of the boundary conditions in an electrochemical system where the electroactive species A is reduced to B which is involved in a chemical reaction in solution that yields the electroinactive species Y.

electrode surface. If the volume of the electrochemical cell is much larger than the region perturbed by the electrode reaction, it can be supposed that there exists an inexhaustible reservoir of species at large distances from the electrode (the bulk solution) such that

$$c_j(x \to \infty, t) = c_j^* \tag{1.17}$$

However, there are some electrochemical experiments where the volume of the liquid phase is comparable to that of the depletion layer such as the case of porous electrode surfaces, microvolumetric cells and amalgamation processes. In these cases, the domain of the solution phase is confined to a distance d from the electrode surface at which no flux of species takes place so that the limit condition is given by

$$D_j \left(\frac{\partial c_j}{\partial x} \right)_{x=d} = 0 \tag{1.18}$$

The second spatial boundary condition relates to the transformation of the species at the electrode surface and it will involve the surface concentration and/or gradient of the corresponding species. In the case of an electroinactive species there will be no flux through the electrode area.

Therefore, similarly to the case of an "electroinactive surface" (Eq. (1.18)) the following condition applies:

$$D_j \left(\frac{\partial c_j}{\partial x} \right)_{x=0} = 0 \qquad (1.19)$$

For the electroactive species, the form of the surface conditions depends on the electrochemical technique employed (see Section 1.2). Thus, in current-controlled techniques, the surface condition establishes the surface flux of the species according to the value of the current imposed (Eq. (1.14)). In potential-controlled methods the applied potential determines the surface concentration and flux of the electroactive species through the kinetics of the electrode reaction. In general, for the one-step process (1.1), this can be expressed according to the following first-order rate law for an interfacial process:

$$D_A \left(\frac{\partial c_A}{\partial x} \right)_{x=0} = k_f c_A(x=0) - k_b c_B(x=0) \qquad (1.20)$$

where k_f and k_b ($\mathrm{m\,s^{-1}}$) are the heterogeneous rate constants of reduction and oxidation, respectively. As discussed in Section 1.1, the value of the rate constants depends on the applied potential such that k_f increases and k_b decreases as the potential takes more negative values. Different kinetic formalisms are available to describe the form of the potential dependence of the rate constants, the Butler–Volmer [11, 12] and Marcus–Hush [13–15] models being the most widely employed. In the limit of fast electrode reactions (so-called reversible processes) both models yield the Nernstian relationship which establishes the equilibrium surface concentrations of the electroactive species:

$$\frac{c_A(x=0)}{c_B(x=0)} = e^{\left(\frac{F}{RT}(E-E_f^0) \right)} \qquad (1.21)$$

Regardless of the technique employed, the conservation of mass principle establishes that the surface flux of the reactive species A must be the same as the surface flux of the product B:

$$D_B \left(\frac{\partial c_B}{\partial x} \right)_{x=0} = \mp D_A \left(\frac{\partial c_A}{\partial x} \right)_{x=0} \qquad (1.22)$$

The upper sign refers to the situation where both species diffuse in the same phase (the most common situation in electrochemical studies) and the lower one in different phases (e.g., amalgamation processes and ion transfers at liquid-liquid interfaces).

1.1.3. *Reaction mechanisms*

Another aspect that must be defined for the system to be simulated is the reaction mechanism, that is, the number and nature of heterogeneous electron transfer reactions (referred to as E steps) and homogeneous processes (C steps) taking place.

The simplest case corresponds to the one-electron transfer between the electrode and species that are chemically stable on the time scale of the experiments (Eq. (1.1)). However, electrochemical systems are frequently more complicated and the electroactive species take part in successive electron transfer reactions at the electrode (multistep processes) and/or in chemical reactions in solution such as protonation, dimerisation, rearrangement, electron exchange, nucleophilic/electrophilic addition, disproportionation, etc., the product(s) of which may or may not be electroactive in the potential region under study. The simulation of these cases is described in Chapters 5 and 6.

Multistep processes can be found in the electroreduction/oxidation of important species such as oxygen, organometallic compounds, biomolecules (e.g., nucleic acids, metalloproteins, enzymes, oligonucleotides), aromatic hydrocarbons and nanoparticles.

$$A^{z_A} + e^- \underset{k_{b,1}}{\overset{k_{f,1}}{\rightleftarrows}} B^{z_A-1} \quad (E^0_{f,1})$$

$$B^{z_A-1} + e^- \underset{k_{b,2}}{\overset{k_{f,2}}{\rightleftarrows}} C^{z_A-2} \quad (E^0_{f,2}) \tag{1.23}$$

$$C^{z_A-2} + e^- \underset{k_{b,3}}{\overset{k_{f,3}}{\rightleftarrows}} D^{z_A-3} \quad (E^0_{f,3})$$

$$\cdots$$

A major interest in this field is the study of the stability of the different oxidation states, which can give information about the structure and bonding of the electroactive species, as well as the medium effects of solvent and dissolved species. In this context, electrochemical techniques are very valuable for the extraction of the E^0_f values.

When the electroactive species are involved in chemical reactions in solution, the electrochemical response is affected by the characteristics of these reactions. Figures 1.8 and 1.9 summarise the most common reaction schemes that can be found.

Note that the homogeneous chemical reactions (C steps) coupled to the electrode process alter the concentration profiles of the electroactive species and therefore the electrochemical response of the system. Thus, electrochemical methods enable the characterisation of the chemical

Fig. 1.8.   Common electrode reaction mechanisms found in voltammetric studies (I). Species A and B are electroactive whereas species Y is assumed to be electroinactive in the potential region of study.

Catalytic mechanism

$$A + e^- \leftrightarrows B$$

$$B + Z \leftrightarrows A + Y$$

dopamine

+ 2 e^- + 2 H^+

+ NADH + H^+ ⟶ + NAD^+

(NAD^+ = nicotinamide adenine dinucleotide)

ECE mechanism   Electroreduction of halonitroaromatic compounds in aprotic media

$$A + e^- \leftrightarrows B$$

$$B \leftrightarrows C$$

$$C + e^- \leftrightarrows D$$

+ e^- ⇌

⟶ + X^-

+ HS ⟶ + S^•

+ e^- ⇌

(X= halogen)
(HS = solvent / supporting electrolyte)

Fig. 1.9. Common electrode reaction mechanisms found in voltammetric studies (II). Species A, B, C and D are electroactive whereas species Y and Z are assumed to be electroinactive in the potential region of study.

reaction in solution, that is, the determination of the mechanism as well as the values of the rate and equilibrium constants.

1.2. Voltammetric Techniques

Once the system to simulate is properly described, we have to consider the electrical perturbation applied, that is, the electrochemical technique employed. This book will focus on the simulation of voltammetric techniques where the potential of the working electrode is varied according to different potential-time programs and the current response of the system is registered. Figure 1.10 outlines the potential-time programs and signals obtained for the main controlled-potential methods.

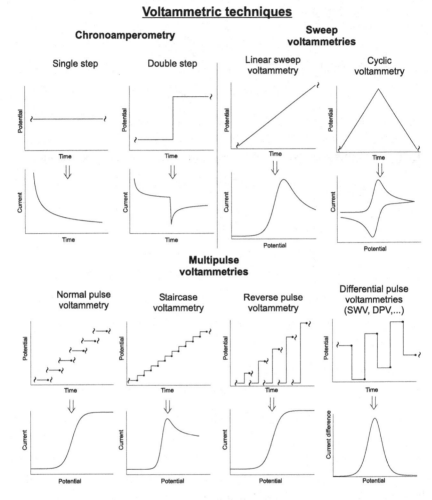

Fig. 1.10. Perturbation (potential-time program) and signal in voltammetric techniques. The sampling time in pulse techniques is marked with a circle and the time where the circuit is open with the symbol ⟨.

In *single-* and *double-step chronoamperometries*, one or two constant potential pulses are applied and the signal corresponds to the variation of the resultant current with time. When a time-varying potential is applied, this can be done linearly (*linear sweep* and *cyclic voltammetries*) or by pulses. In any case, the current measured is plotted versus the corresponding potential.

Linear sweep and cyclic voltammetries are probably the most widely used techniques given that they enable simple and direct qualitative analysis of the electrochemical system. However, caution must be taken since the program applied with modern digital equipment does not correspond to a linear ramp but rather to a staircase.

Regarding pulse voltammetries [16–18], a notable variety of methods have been developed where the current is usually sampled at the end of each potential pulse in order to minimise the non-faradaic current from the charging of the electrical double layer at the electrode/solution interface. This group of techniques includes *staircase voltammetry, reverse pulse voltammetry* and differential pulse voltammetries. Differential methods such as *square wave* and *differential pulse voltammetries* are extensively employed in electroanalysis since they offer a greater sensitivity and signals more appropriate for quantitative studies. In these techniques, the current of two consecutive pulses is subtracted which gives rise to the reduction of background currents and to well-defined, peaked voltammograms.

1.3. Finite Difference Methods

Once the electrochemical system and the perturbation applied are properly defined, there exist different numerical methods to solve the corresponding differential equation problem. Finite difference methods will be employed in this book given that in all the situations considered here they provide simple, accurate and effective procedures.[8]

As will be detailed in the following chapters, in these methods the differential equations describing the mass transport are approximated as finite difference equations. This requires the discretisation of the domains of the independent variables (time and distance) by equally or unequally spaced grids. The distribution of nodes in the grids must ensure the accuracy of the finite difference approximation. Thus, the number of points must be high enough in the time and space regions where the changes in concentration are more significant (typically near the electrode surface).

Several procedures for the resolution of the resulting difference equation systems will be discussed together with criteria to check the accuracy of the numerical results.

In the next chapter the simulation of cyclic voltammetry at macro-electrodes will be introduced. The main aspects of the most widely used

---

[8] For details about other numerical methods, see Britz and Strutwolf's book [19].

voltammetric technique will be described as well as the mathematical treatment of the differential equation system and boundary conditions prior to the resolution of the problem.

1.4. Voltammetry: A Selected Bibliography

In this first chapter some fundamental aspects of electrochemical processes and experiments have been summarised to assist the comprehension of the following chapters. For a deeper analysis of the principles of electrochemistry the following books are recommended:

- Bard, A. J. and Faulkner, L. R. (2001). *Electrochemical Methods: Fundamentals and Applications*, 2nd ed., John Wiley and Sons, New York.

- Bond, A. M. (2002). *Broadening Electrochemical Horizons*, Oxford University Press, Oxford.

- Compton, R. G. and Banks, C. E. (2018). *Understanding Voltammetry*, 3rd ed., World Scientific Press, Singapore.

- Compton, R. G., Batchelor-McAuley, C. and Dickinson, E. J. F. (2012). *Understanding Voltammetry: Problems and Solutions*, Imperial College Press, London.

- Girault, H. H. (2004). *Analytical and Physical Electrochemistry*, EPFL-Press, Lausanne.

- Hammerich, O. and Speiser, B. (2015). *Organic Electrochemistry: Revised and Expanded*, CRC Press.

- Molina, A. and González, J. (2016). *Pulse Voltammetry in Physical Electrochemistry and Electroanalysis*, Springer, Berlin.

- Oldham, K. B., Myland, J. C. and Bond, A. M. (2012). *Electrochemical Science and Technology*, Wiley, Chichester.

- Pletcher, D. (2009). *A First Course in Electrode Processes*, 2nd ed., The Royal Society of Chemistry, Cambridge.

- Savéant, J.-M. (2006). *Elements of Molecular and Biomolecular*

Electrochemistry: An Electrochemical Approach to Electron Transfer Chemistry, Wiley-Interscience, Hoboken.

- Wang, J. (2000). *Analytical Electrochemistry*, 2nd ed., John Wiley and Sons, New York.

References

[1] H. Gerischer, Kinetics of oxidation-reduction reactions on metals and semiconductors. I and II General remarks on the electron transition between a solid body and a reduction-oxidation electrolyte, *Z. Phys. Chem.* **26**, 223–247, (1960).

[2] J. Crank, *The Mathematics of Diffusion.* (Clarendon, Oxford, 1976).

[3] A. Fick, On liquid diffusion, *Lond. Edinb. Dubl. Phil. Mag.* **10**, 30–39, (1855).

[4] A. Fick, Ueber diffusion, *Ann. Phys. (Berlin).* **94**, 59–86, (1855).

[5] C. Batchelor-McAuley, E. J. F. Dickinson, N. Rees, K. E. Toghill, and R. G. Compton, New electrochemical methods, *Anal. Chem.* **84**, 669–684, (2012).

[6] F. G. Cottrell, Der Reststrom bei galvanischer Polarisation, betrachtet als ein Diffusionsproblem, *Z. Phys. Chem.* **42**, 385–431, (1903).

[7] S. R. Belding, J. G. Limon-Petersen, E. J. F. Dickinson, and R. G. Compton, Voltammetry in the absence of excess supporting electrolyte offers extra kinetic and mechanistic insights: Comproportionation of anthraquinone and the anthraquinone dianion in acetonitrile, *Angew. Chem. Int. Ed.* **49**, 9242–9245, (2010).

[8] E. J. F. Dickinson, J. G. Limon-Petersen, N. V. Rees, and R. G. Compton, How much electrolyte is required to make a cyclic voltammetry experiment quantitatively "diffusional"? A theoretical and experimental investigation, *J. Phys. Chem. C.* **113**, 11157–11171, (2009).

[9] C. Amatore, C. Pebay, L. Thouin, A. Wang, and J.-S. Warkocz, Difference between ultramicroelectrodes and microelectrodes: Influence of natural convection, *Anal. Chem.* **82**, 6933–6939, (2010).

[10] J. K. Novev and R. G. Compton, Natural convection effects in electrochemical systems, *Current Opinion in Electrochemistry.* **7**, 118–129, (2018).

[11] J. A. V. Butler, Studies in heterogeneous equilibria. Part II — The kinetic interpretation of the nernst theory of electromotive force, *Trans. Faraday Soc.* **19**, 729–733, (1924).

[12] T. Erdey-Gruz and M. Volmer, Zur Theorie der Wasserstoffüberspannung, *Z. Phys. Chem.* **150A**, 203–211, (1930).

[13] R. A. Marcus, On the theory of oxidation-reduction reactions involving electron transfer. I, *J. Chem. Phys.* **24**, 966–978, (1956).

[14] N. S. Hush, Adiabatic rate processes at electrodes. I. Energy-charge relationships, *J. Chem. Phys.* **28**, 962–972, (1958).

[15] C. E. D. Chidsey, Free energy and temperature dependence of electron transfer at the metal-electrolyte interface, *Science.* **251**, 919–922, (1991).

[16] V. Mirčeski, S. Komorsky-Lovrić, and M. Lovric, *Square-wave voltammetry: Theory and application.* (Springer, Berlin, 2007).

[17] A. Molina, E. Laborda, F. Martínez-Ortiz, D. F. Bradley, D. J. Schiffrin, and R. G. Compton, Comparison between double pulse and multipulse differential techniques, *J. Electroanal. Chem.* **659**, 12–24, (2011).

[18] C. Serna, M. M. Moreno, A. Molina, and J. González, A general survey of the problem of multipotential step techniques at spherical electrodes when both oxidized and reduced species are initially present, *Recent Res. Devel. Electrochem.* **3**, 29–42, (2000).

[19] D. Britz and J. Strutwolf, *Digital Simulation in Electrochemistry. 4th ed.* (Springer, Berlin, 2016).

Chapter 2

Mathematical Model of an Electrochemical System

2.1. Cyclic Voltammetry

The popularity of the cyclic voltammetry (CV) technique has led to its extensive study and numerous simple criteria are available for immediate analysis of electrochemical systems from the shape, position and time-behaviour of the experimental voltammograms [1, 2]. For example, a quick inspection of the cyclic voltammograms offers information about the diffusive or adsorptive nature of the electrode process, its kinetic and thermodynamic parameters, as well as of the existence and characteristics of coupled homogeneous chemical reactions [2]. This electrochemical method is also very useful for the evaluation of the magnitude of undesirable effects such as those derived from ohmic drop or double-layer capacitance. Accordingly, cyclic voltammetry is frequently used for the analysis of electroactive species and surfaces, and for the determination of reaction mechanisms and rate constants.

We can represent a general reduction process at the electrode as

$$A + e^- \rightleftharpoons B \tag{2.1}$$

In a typical cyclic voltammetry experiment, the potential is swept linearly with time from some starting potential, E_i, where species A is stable (i.e., not electroreduced), to some other, more negative potential, E_v, at which electron transfer between species A and the electrode is rapid, and species B is formed. The potential is then swept back to E_i, causing electron transfer in the opposite direction and the reformation of A. This potential waveform is shown in Figure 2.1. Throughout this process the current, I (proportional to the rate of electron transfer), is recorded; plotting the current against the potential gives a characteristic peaked cyclic voltammogram of a macroelectrode as shown in Figure 2.2.

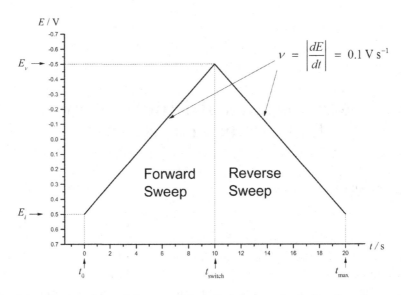

Fig. 2.1. The waveform of the potential applied during a typical cyclic voltammetry experiment. In this case the initial potential, E_i, is 0.5 V, the vertex potential, E_v, is -0.5 V, and the scan rate, ν, is 0.1 V s^{-1}.

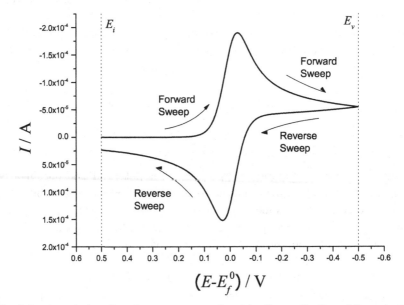

Fig. 2.2. A typical cyclic voltammogram produced by the application of the potential waveform in Figure 2.1.

This technique is extremely useful experimentally as the resulting peak-shaped signal provides a direct fingerprint of the features of the reduction and oxidation processes. Analysis of the position and shape of the peaks can give important information about the nature of the electrochemical process taking place and about the chemical species themselves.

The scan rate, ν, in V s^{-1}, is the constant rate at which the voltage sweeps from the initial potential, E_i, to the vertex potential, E_v, and back again. Figure 2.1 demonstrates that the scan rate is the rate of change of potential (the slope) defined as

$$\nu = \left| \frac{dE}{dt} \right| \tag{2.2}$$

At any time t on the forward sweep, the potential, E, is given by

$$E = E_i - \nu t \tag{2.3}$$

At time $t = t_{\text{switch}}$, the potential reaches E_v and the potential sweep reverses direction. For $t > t_{\text{switch}}$,

$$E = E_v + \nu(t - t_{\text{switch}}) \tag{2.4}$$

or equivalently

$$E = 2E_v - E_i + \nu t \tag{2.5}$$

since

$$t_{\text{switch}} = \left| \frac{E_i - E_v}{\nu} \right| \tag{2.6}$$

2.2. Diffusion: Fick's Second Law

Let us suppose an electrochemical experiment in which at the beginning, the concentration of the chemical species, A, is uniform throughout the entire solution and there is no B present at all. When a negative enough potential is applied, A is converted into B at the electrode, thus in the vicinity of the electrode, the concentration of A decreases and that of B increases. Consequently, fresh A diffuses toward the electrode from the bulk solution, and B diffuses away from the electrode. This process is illustrated schematically in Figure 2.3.

A basic understanding of the principles of diffusion and electrode kinetics allows us to explain the shape of the cyclic voltammogram (Figure 2.2).

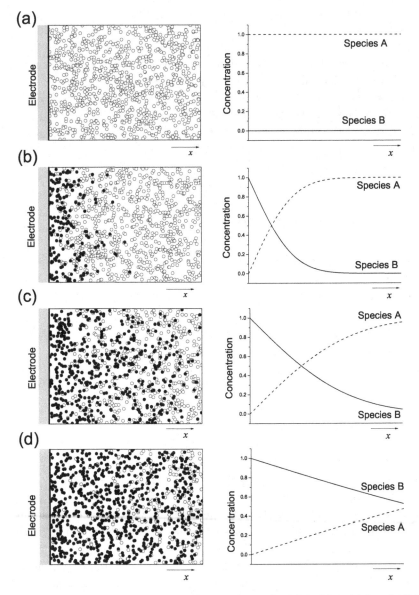

Fig. 2.3.   Schematic showing the distribution of particles (a) 0, (b) 1, (c) 5 and (d) 50 arbitrary time units after a potential pulse is applied to the electrode. White dots are the starting species, A, and black dots are the reduced species, B. Concentration profiles over the same space are also shown.

At the start of the scan, when we are at a potential that is positive relative to the formal potential of the redox couple, no current flows because the potential is not negative enough to reduce A to B. As the scan progresses and the potential is made more negative, the reduction reaction becomes progressively more thermodynamically favourable and so the current increases with increasing potential (this explains the rising section of the forward sweep). Eventually the concentration of the reactant species A in the immediate vicinity of the electrode drops to zero and the current reaches a maximum. Beyond this point, the current decays as the reaction is now under *diffusional control*; the rate of electron transfer is limited by the rate at which fresh material can be brought to the electrode by diffusion even though the reaction continues to become progressively more favourable thermodynamically.

Fick's second law [3–5] predicts how the diffusion causes the concentration field to change with time:

$$\frac{\partial c}{\partial t} = D \left(\frac{\partial^2 c}{\partial x^2} + \frac{\partial^2 c}{\partial y^2} + \frac{\partial^2 c}{\partial z^2} \right) \tag{2.7}$$

where the diffusion coefficient, D, is assumed to be direction-independent and c is the concentration of the species which is a function of the time, t, and spatial coordinates, x, y and z. The diffusion coefficient, D, is particular to each chemical species and is a measure of the rate at which the species moves through solution; a typical value is 10^{-5} cm^2 s^{-1}. Note that each chemical species is modelled by its own version of this equation, with its own concentration values, e.g., c_A and c_B, and diffusion coefficients D_A and D_B. Simple electrochemical simulation consists of solving this partial differential equation (PDE) subject to certain boundary conditions, thereby determining the full time evolution of the system.

Modelling a three-dimensional PDE is extremely time consuming, even for a fast computer, so we use a simplification. If we use an electrode that is large relative to the distance over which the species diffuse on the time scale of the experiment, that is, a macroelectrode, then we only need to consider diffusion that is perpendicular to the plane of the electrode surface as there will be negligible diffusion parallel to the surface as shown in Figure 2.4(a). The problem is thus reduced to one spatial dimension, x, which is the distance normal to the surface of the electrode. For our purposes, we will assume that the electrode is a disc shape with radius ϵ as shown in Figure 2.4(b). In general, however, the shape is not important, as long as the electrode is flat and large in comparison to the thickness of

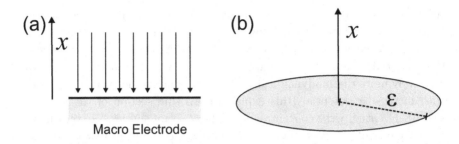

Fig. 2.4.   (a) The diffusion field at a planar macroelectrode; (b) a one-dimensional macrodisc electrode with radius ϵ and normal coordinate x.

the diffusion layer, δ, where

$$\delta \sim \sqrt{Dt} \qquad (2.8)$$

For a typical experimental setup with $D = 10^{-5}$ cm^2 s^{-1} and a running time, t, of 10 s, this gives an order of magnitude estimate of $\delta = 0.1$ mm. Therefore an electrode is typically considered to be of macroscale if it is on the order of 1 mm in size or greater.

As the problem has been reduced to one spatial dimension, Eq. (2.7) is simplified considerably and diffusion in this new coordinate system may be modelled with the equation

$$\frac{\partial c}{\partial t} = D \left(\frac{\partial^2 c}{\partial x^2} \right) \qquad (2.9)$$

2.2.1. *Equal diffusion coefficients*

Fick's first law states that the diffusive flux, j_{x_1}, through a plane normal to the x-axis at the point in space x_1 is equal to the product of the diffusion coefficient and the concentration gradient at that point:

$$j_{x_1} = -D \left(\frac{\partial c}{\partial x} \right)_{x_1} \qquad (2.10)$$

where the minus sign implies that the flux is down the local concentration gradient.

At the start of an experiment, the solution contains just species A at a uniform concentration, and A is transformed into B at the electrode surface. By conservation of mass, the amount of A lost at the electrode surface exactly equals the amount of B gained there, i.e., $-\Delta c_A = \Delta c_B$, and therefore the diffusive fluxes of the two species at the electrode surface are equal and opposite:

$$D_B \left(\frac{\partial c_B}{\partial x} \right)_{\text{electrode}} = -D_A \left(\frac{\partial c_A}{\partial x} \right)_{\text{electrode}} \qquad (2.11)$$

If we assume that the diffusion coefficients of both species are equal, i.e., $D = D_A = D_B$, then a useful simplification may be made. It follows that

$$\left(\frac{\partial c_B}{\partial x} \right)_{\text{electrode}} = - \left(\frac{\partial c_A}{\partial x} \right)_{\text{electrode}} \qquad (2.12)$$

therefore a decrease in concentration of species A at some point in space is necessarily commensurate with an increase in the concentration of species B of the same magnitude so that

$$\frac{\partial (c_A + c_B)}{\partial x} = 0 \qquad (2.13)$$

Therefore it is necessarily true that in every region of space

$$c_A = (c_A^* + c_B^*) - c_B \qquad (2.14)$$

where c_A^* and c_B^* are the initial concentrations of species A and B respectively (see below). Consequently, when modelling our electrochemical system, we only need to consider the diffusional behaviour of species A, since we can infer the behaviour of B directly from that of A.

2.3. Boundary Conditions

Equation (2.9) is an example of a parabolic second-order partial differential equation. The equation describes a single property, concentration, which evolves in space and time. In order to solve an equation of this type, we need to know the condition of the system at some starting time, $t = 0$. We have already stated that at the start of the experiment, the concentration of species A is a fixed value (1 mM for example) and is uniform everywhere. We call this the bulk concentration of species A and represent it with the symbol c_A^*. Therefore we have the *initial condition*:

$$c_A(x, t = 0) = c_A^* \qquad (2.15)$$

A similar condition applies to species B, except that initially we are going to assume that there is no B present, so $c_B^* = 0$. At times $t > 0$, the evolution of the system is given by Fick's second law (Eq. (2.9)).

As it stands, our system is not very interesting; it is infinite in extent and as the concentration of both species is initially uniform, there are no concentration gradients and hence no diffusion, so the system will be unchanging in time. In order to solve a particular electrochemical problem, we must impose some *spatial boundary conditions* on the concentration. Our one-dimensional space has two spatial boundaries; we therefore constrain the system to some finite region:

$$x_0 \leq x \leq x_{max} \tag{2.16}$$

where x_0 and x_{max} are the lower and upper spatial bounds respectively. We then impose a boundary condition on the concentration at each of these boundaries. These may alter the concentrations of the chemical species in some way such that a concentration gradient is established and diffusion occurs.

The first boundary is the surface of the macrodisc electrode, which we define to be at $x = 0$. When a potential difference is applied across the electrode, electron transfer occurs, transforming species A at the surface into species B. Therefore, the concentrations at the electrode surface vary as a function, f, of the potential, E, applied to the electrode. In general, we can write

$$c(x = 0, t > 0) = f(E) \tag{2.17}$$

Assuming that the kinetics of the electron transfer are fast relative to the rate of mass transport, Nernstian equilibrium is attained at the electrode surface throughout the potential scan, and the Nernst equation therefore relates the surface concentrations of species A and B to the applied potential at the electrode, E:

$$E = E_f^0 + \frac{\mathcal{R}\mathcal{T}}{F} \ln\left(\frac{c_{A,0}}{c_{B,0}}\right) \tag{2.18}$$

where E_f^0 is the formal potential of the reaction, $\mathcal{R}$ is the gas constant, $\mathcal{T}$ is the temperature (often 298 K), F is the Faraday constant, and $c_{A,0}$ and $c_{B,0}$ are the concentrations at (next to) the electrode surface of A and B respectively. The Butler–Volmer and Marcus–Hush models which take electron transfer kinetics into account will be examined in Chapter 4.

The upper spatial boundary may be defined in a number of ways. Ideally we would define it as being infinitely far away from the electrode, i.e., $x_{\max} \to +\infty$, such that changes in concentration at the electrode cannot have any effect on the concentration at the upper boundary on the time scale of the experiment. In practice, it transpires that it is not necessary to place the boundary infinitely far away from the electrode in order to meet this condition. Einstein's work on Brownian motion in 1905 [6] demonstrated that in one dimension, the root mean squared displacement of a particle from its starting position, $\sqrt{x^2}$ (alternatively, $\bar{x}$), is equal to

$$\bar{x} = \sqrt{2Dt} \tag{2.19}$$

To be sure that no significant changes in concentration can propagate as far as the boundary, we must set it to be at some distance greater than $\bar{x}$. A typical value is

$$x_{\max} = 6\sqrt{Dt_{\max}} \tag{2.20}$$

where $t_{\max}$ is the maximum time the experiment will run for. For a typical diffusion coefficient of 10^{-5} cm^2 s^{-1} and experimental time of 10 s, this distance is only 0.6 mm. For our purposes, a distance of $x_{\max}$ given by Eq. (2.20) may reasonably be defined as being infinitely far away from the electrode. Since the effects of diffusion will not extend this far from the electrode during the experiment, the concentration of each species at this distance does not change with time, and so is always equal to its bulk value:

$$c(x = x_{\max}, t > 0) = c^* \tag{2.21}$$

Figure 2.5 summarises this mathematical model of the system. Note that typically, an electrochemical experiment will take place inside some enclosure, e.g., test tube, beaker, so the size of the experimental space may be constrained in this way; however, the distance from the electrode surface to the wall of the container is likely to be on the order of a centimetre, greatly exceeding the value of $x_{\max}$.

Boundary conditions in which the values of the boundary points are specified either as a constant or as a function of time are known as *Dirichlet boundary conditions*. Later we will also find *Neumann boundary conditions*, where the value of the first derivative of concentration (the concentration gradient) across the boundary is specified, as well as *mixed conditions*.

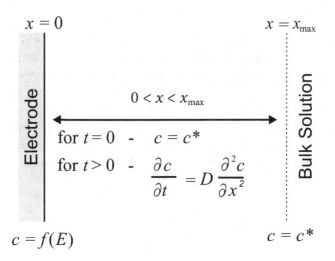

Fig. 2.5.   The one-dimensional space described by our model with boundary conditions.

2.4. Current

The primary output of a voltammetry experiment or simulation is the voltammogram which is a plot of recorded current, I, against applied potential, E. At any point in time, the current for the reduction of species A at uniformly accessible electrodes, where the surface flux of redox species is the same at any point on the surface, is given by

$$I = FAj_{0,A} \qquad (2.22)$$

where A is the area of the electrode ($A = \pi\epsilon^2$ for a macrodisc) and $j_{0,A}$ is the net flux of electroactive material (species A) at the electrode surface which is proportional to the concentration gradient at that point by Fick's first law, Eq. (2.10). So in order to calculate the current, we need to find j at the electrode surface ($x = 0$) as it evolves in time.

2.5. Dimensionless Parameters

The particular solution of Fick's second law (Eq. (2.9)) subject to the above boundary conditions changes for all values of bulk concentration, c^*, electrode radius, ϵ, and diffusion coefficient, D, and hence must be solved again when any one of these values change. We can transform our system variables (c, x, t, etc.) into a dimensionless form by expressing each of them as multiples of some characteristic reference value with the same dimensionality. This removes the dependence of the solution of Fick's second law on c^*, ϵ and D, making it much more general such that the equation needs only to be solved once; the resulting voltammogram can then simply be scaled to apply to any set of values of c^*, ϵ and D.

Here we present a set of suitable coordinate transformations. This is by no means the only sensible set of transformations; the characteristic reference value for each variable should be chosen appropriately according to the particulars of the system under investigation. First we define the dimensionless concentration of chemical species j, C_j, to be

$$C_j = \frac{c_j}{c_A^*} \qquad (2.23)$$

where $j \equiv A$ or B. The dimensionless concentration is the real concentration divided by the real bulk concentration of species A. Therefore, whether the bulk concentration of A is 1 mM or 100 mM, the dimensionless concentration of A is always initially unity, i.e., $C_A = 1$. Remember also that since the diffusion coefficients are equal, from Eq. (2.14) it is necessarily true that $C_A + C_B = 1$ at every point in space (since it is assumed that $c_B^* = 0$). The dimensionless diffusion coefficient, d, of each species is likewise defined:

$$d_j = \frac{D_j}{D_A} \qquad (2.24)$$

If the diffusion coefficients are equal then $d_A = d_B = 1$. Later we will examine the case where species diffuse at different rates, which is why we define d as a ratio of real diffusion coefficients rather than simply as 1.

We define the dimensionless distance X in terms of the radius of the macrodisc electrode, ϵ:

$$X = \frac{x}{\epsilon} \qquad (2.25)$$

such that the dimensionless radius of the electrode is 1. If the electrode is not a disc shape then some other characteristic length, such as the side length for a square electrode, may be used for ϵ. Finally we define the dimensionless time, T, in terms of both the radius and the diffusion coefficient of species A:

$$T = \frac{D_A t}{\epsilon^2} \qquad (2.26)$$

Notice that the quantities C, d, X and T all have no units.

It is now necessary to transform Eq. (2.9) and the boundary conditions into this new dimensionless coordinate system. We begin by considering the left-hand side of Eq. (2.9), $\partial c/\partial t$. We may make the substitution, $c = c_A^* C$ (Eq. (2.23)), using the chain rule

$$\frac{\partial y}{\partial x} = \frac{\partial u}{\partial x}\frac{\partial y}{\partial u} \tag{2.27}$$

or specifically in this case with $j = A$ or B

$$\frac{\partial c_j}{\partial t} = \frac{\partial C_j}{\partial t}\frac{\partial c_j}{\partial C_j} \tag{2.28}$$

Now since

$$\frac{\partial c_j}{\partial C_j} = c_A^* \tag{2.29}$$

we are left with

$$\frac{\partial c_j}{\partial t} = c_A^* \frac{\partial C_j}{\partial t} \tag{2.30}$$

We can similarly make the substitution $t = T\epsilon^2/D_A$ (Eq. (2.26)) in our new partial derivative $\partial C_j/\partial t$:

$$\frac{\partial C_j}{\partial t} = \frac{\partial C_j}{\partial T}\frac{\partial T}{\partial t} \tag{2.31}$$

And since

$$\frac{\partial T}{\partial t} = \frac{D_A}{\epsilon^2} \tag{2.32}$$

we have

$$\frac{\partial C_j}{\partial t} = \frac{D_A}{\epsilon^2}\frac{\partial C_j}{\partial T} \tag{2.33}$$

Consequently

$$\frac{\partial c_j}{\partial t} = \frac{c_A^* D_A}{\epsilon^2}\frac{\partial C_j}{\partial T} \tag{2.34}$$

This might not seem particularly useful, but let us consider the right-hand side of Eq. (2.9) using the chain rule for the second derivative:

$$\frac{\partial^2 y}{\partial x^2} = \frac{\partial}{\partial x}\left[\frac{\partial u}{\partial x}\frac{\partial y}{\partial u}\right] = \left[\frac{\partial u}{\partial x}\right]^2\frac{\partial^2 y}{\partial u^2} + \left[\frac{\partial^2 u}{\partial x^2}\right]\frac{\partial y}{\partial u} \tag{2.35}$$

First we make the substitution $c = c_A^* C$:

$$\frac{\partial^2 c_j}{\partial x^2} = \left[\frac{\partial C_j}{\partial x}\right]^2 \frac{\partial^2 c_j}{\partial C_j^2} + \left[\frac{\partial^2 C_j}{\partial x^2}\right] \frac{\partial c_j}{\partial C_j} \tag{2.36}$$

$$= c_A^* \frac{\partial^2 C_j}{\partial x^2} \tag{2.37}$$

since

$$\frac{\partial^2 c_j}{\partial C_j^2} = 0 \tag{2.38}$$

Next we make the substitution, $x = \epsilon X$ (Eq. (2.25)):

$$\frac{\partial^2 C_j}{\partial x^2} = \left[\frac{\partial X}{\partial x}\right]^2 \frac{\partial^2 C_j}{\partial X^2} + \left[\frac{\partial^2 X}{\partial x^2}\right] \frac{\partial C_j}{\partial X} \tag{2.39}$$

$$= \frac{1}{\epsilon^2} \frac{\partial^2 C_j}{\partial X^2} \tag{2.40}$$

since

$$\frac{\partial^2 X}{\partial x^2} = 0 \tag{2.41}$$

Consequently

$$D_j \frac{\partial^2 c_j}{\partial x^2} = d_j D_A \frac{c_A^*}{\epsilon^2} \frac{\partial^2 C_j}{\partial X^2} \tag{2.42}$$

Therefore we may write Eq. (2.9) in terms of our dimensionless parameters, C, X and T:

$$\frac{c_A^* D_A}{\epsilon^2} \frac{\partial C_j}{\partial T} = d_j \frac{c_A^* D_A}{\epsilon^2} \frac{\partial^2 C_j}{\partial X^2} \tag{2.43}$$

which simplifies to

$$\frac{\partial C_j}{\partial T} = d_j \frac{\partial^2 C_j}{\partial X^2} \tag{2.44}$$

Note that every quantity in Eq. (2.44) is dimensionless. We must now also transform the boundary conditions into a dimensionless form:

$$c_j (x, t = 0) = c_j^* \rightarrow \begin{cases} C_A (X, T = 0) = 1 \\ C_B (X, T = 0) = 0 \end{cases}$$

$$c_j (x = x_{\text{max}}, t) = c_j^* \rightarrow \begin{cases} C_A (X = X_{\text{max}}, T) = 1 \\ C_B (X = X_{\text{max}}, T) = 0 \end{cases} \tag{2.45}$$

where

$$X_{\max} = 6\sqrt{T_{\max}} \tag{2.46}$$

2.5.1. *Electrode surface boundary*

By rearranging the Nernst equation (Eq. (2.18)), we get

$$(E - E_f^0)\frac{F}{RT} = \ln\left(\frac{c_{A,0}}{c_{B,0}}\right) \tag{2.47}$$

We define the dimensionless potential, θ, to be

$$\theta = \frac{F}{RT}(E - E_f^0) \tag{2.48}$$

so that

$$e^\theta = \frac{c_{A,0}}{c_{B,0}} \tag{2.49}$$

and since $c_{A,0}/c_{B,0} = C_{A,0}/C_{B,0}$

$$e^\theta = \frac{C_{A,0}}{C_{B,0}} \tag{2.50}$$

Now if $C_A + C_B = 1$ (from Eq. (2.14))

$$e^\theta = \frac{C_{A,0}}{1 - C_{A,0}} \tag{2.51}$$

and so we have the boundary condition for concentration at the electrode surface:

$$\begin{aligned}
C_A(X = 0, T) &= \frac{1}{1 + e^{-\theta}} \\
C_B(X = 0, T) &= \frac{1}{1 + e^{\theta}}
\end{aligned} \tag{2.52}$$

where θ is typically a function of time. Figure 2.6 shows how the concentrations of species A and B at the surface of the electrode vary with θ. Note that when $T = 298$ K, one unit of θ is approximately 25.7 mV.

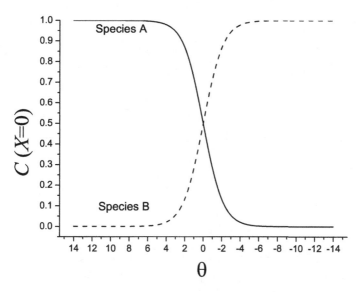

Fig. 2.6. Equilibrium surface concentrations of species A and B as a function of θ.

2.5.2. *Scan rate*

As shown in Eq. (2.2), the scan rate is the rate of change of potential with respect to time. We can therefore define the dimensionless scan rate, σ, to be the rate of change of dimensionless potential with respect to dimensionless time:

$$\sigma = \frac{\partial \theta}{\partial T} \tag{2.53}$$

We therefore need to determine how σ relates to ν. We again use the chain rule:

$$\frac{\partial E}{\partial t} = \frac{\partial E}{\partial \theta} \frac{\partial \theta}{\partial t} = \frac{\mathcal{R} \mathcal{T}}{\mathrm{F}} \frac{\partial \theta}{\partial T} \tag{2.54}$$

since (from Eq. (2.48)):

$$\frac{\partial E}{\partial \theta} = \frac{\mathcal{R} \mathcal{T}}{\mathrm{F}} \tag{2.55}$$

Then

$$\frac{\partial \theta}{\partial t} = \frac{\partial \theta}{\partial T} \frac{\partial T}{\partial t} = \frac{D_A}{\epsilon^2} \frac{\partial \theta}{\partial T} \tag{2.56}$$

which leads to

$$\frac{\partial E}{\partial t} = \frac{D_A}{\epsilon^2} \frac{\mathcal{R}\mathcal{T}}{F} \frac{\partial \theta}{\partial T} \tag{2.57}$$

Therefore

$$\sigma = \frac{\epsilon^2}{D_A} \frac{F}{\mathcal{R}\mathcal{T}} \nu \tag{2.58}$$

From this we can see that at dimensionless time T in the forward scan, the dimensionless potential is given by

$$\theta = \theta_i - \sigma T \tag{2.59}$$

and for the reverse scan by

$$\theta = 2\theta_v - \theta_i + \sigma T \tag{2.60}$$

where θ_i is the initial dimensionless potential defined as:

$$\theta_i = \frac{F}{\mathcal{R}\mathcal{T}}(E_i - E_f^0) \tag{2.61}$$

and θ_v is likewise defined.

2.5.3. *Current*

Since our model system is in dimensionless units, we will not find the flux, j, directly, but the dimensionless flux at point X_1, J_{X_1}, defined as

$$J_{X_1,j} = -d_j \left(\frac{\partial C_j}{\partial X} \right)_{X_1} \tag{2.62}$$

Using the chain rule again we can determine how J relates to j:

$$\frac{\partial c_j}{\partial x} = \frac{\partial c_j}{\partial X} \frac{\partial X}{\partial x} = \frac{1}{\epsilon} \frac{\partial c_j}{\partial X} \tag{2.63}$$

$$\frac{\partial c_j}{\partial X} = \frac{\partial c_j}{\partial C_j} \frac{\partial C_j}{\partial X} = c_A^* \frac{\partial C_j}{\partial X} \tag{2.64}$$

Therefore

$$\frac{\partial c_j}{\partial x} = \frac{c_A^*}{\epsilon} \frac{\partial C_j}{\partial X} \tag{2.65}$$

$$j_j = -d_j \frac{c_A^* D_A}{\epsilon} \frac{\partial C}{\partial X} \tag{2.66}$$

$$j_j = \frac{c_A^* D_A}{\epsilon} J_j \tag{2.67}$$

So, substituting for j in Eq. (2.22), we get

$$I = \frac{FAc_A^* D_A}{\epsilon} J_{0,A} \tag{2.68}$$

where

$$J_{0,A} = -\left(\frac{\partial C_A}{\partial X}\right)_{X=0} \tag{2.69}$$

When we run a simulation, we record the dimensionless flux, J, as a function of the dimensionless potential, θ. To obtain a real voltammogram, the results of this simulation are simply scaled by the real values of c_A^*, D_A and ϵ. As a consequence, the results of a single simulation will apply to any set of these three values, so we do not need to run another simulation each time we change, for example, the concentration. This is exactly why we introduced dimensionless variables in the first place.

Note that the definition (2.68) holds for uniformly accessible electrodes taking the corresponding characteristic distance (ϵ) employed in the normalisation of the problem, for example, the radius of the electrode in the case of cylinder and (hemi)spherical electrodes (see Chapter 4). In the case of non-uniformly accessible electrodes such as microdiscs (see Chapter 9), where the flux of electroactive species differs at different regions of the electrode surface (two-dimensional systems), the relationship (2.22) does not apply and the definition of the dimensionless current J can be different from (2.68).

2.6. Summary

In this chapter we have established a mathematical model that fully describes a cyclic voltammetry experiment of a one-electron reversible process at a planar macrodisc electrode where the diffusion coefficients of both chemical species are equal. The model consists of a one-dimensional partial differential equation that describes the evolution of the concentration of some chemical species in both time and space starting from some initial conditions at time $t = 0$, and bounded by some finite spatial region $0 \leq x \leq x_{max}$. At $x = 0$ is the electrode boundary which alters the concentration in a manner that depends on the potential applied to it. At $x = x_{max}$ the concentration is unaffected by the processes occurring at the electrode and so is equal to that of the bulk solution. The potential at the

Table 2.1.   Dimensionless parameters and variables.

Parameter	Normalisation
concentration	$C_j = c_j/c_A^*$
diffusion coefficient	$d_j = D_j/D_A$
spatial coordinate	$X = x/\epsilon$
time	$T = D_A t/\epsilon^2$
potential	$\theta = (F/\mathcal{R}T)(E - E_f^0)$
scan rate	$\sigma = (\epsilon^2/D_A)(F/\mathcal{R}T)\nu$
heterogeneous rate constant	$K_0 = k_0\epsilon/D_A$ (Chapter 4)
1st-order homogeneous rate constant	$K = k\epsilon^2/D_A$ (Chapter 5)
2nd-order homogeneous rate constant	$K^{2\mathrm{nd}} = kc_A^*\epsilon^2/D_A$ (Chapter 6)
current (uniformly accessible electrode)	$I = (FAc_A^* D_A/\epsilon)J_{0,A}$

electrode is varied and the resultant current is recorded and plotted as a voltammogram.

For convenience, we normalised this model, reducing it to a set of dimensionless parameters as summarised in Table 2.1. Our model is therefore fully described by the equation

$$\frac{\partial C_j}{\partial T} = d_j \left(\frac{\partial^2 C_j}{\partial X^2} \right) \tag{2.70}$$

subject to the boundary conditions

$$C_A(X, T = 0) = 1 , \; C_B(X, T = 0) = 0 \tag{2.71}$$

$$C_A(X = X_{\max}, T) = 1 , \; C_B(X = X_{\max}, T) = 0 \tag{2.72}$$

$$C_A(X = 0, T > 0) = \frac{1}{1 + e^{-\theta}} , \; C_B(X = 0, T > 0) = \frac{1}{1 + e^{\theta}} \tag{2.73}$$

where

$$X_{\max} = 6\sqrt{T_{\max}} \tag{2.74}$$

$$\theta = \theta_i - \sigma T \qquad \qquad \text{for} \quad T \leq T_{\text{switch}} \tag{2.75}$$

$$\theta = \theta_v + \sigma(T - T_{\text{switch}}) \qquad \text{for} \quad T > T_{\text{switch}} \tag{2.76}$$

To obtain a dimensionless voltammogram, we then record the dimensionless flux of species A, J_A, at the electrode surface as it varies with the dimensionless potential, θ:

$$J_{0,A} = -\left(\frac{\partial C_A}{\partial X}\right)_0 \tag{2.77}$$

From this, a real voltammogram in terms of real current I and real potential E is produced simply by scaling $J_{0,A}$ and θ according to

$$I = \frac{F A c_A^* D_A}{\epsilon} J_{0,A} \tag{2.78}$$

$$E = E_f^0 + \frac{\mathcal{R}\mathcal{T}}{F}\theta \tag{2.79}$$

In the next chapter we will develop more mathematical theory that will allow us to solve this model using a computer.

References

[1] A. J. Bard and L. R. Faulkner, *Electrochemical Methods: Fundamentals and Applications.* (John Wiley & Sons, New York, 2001), 2nd edition.

[2] R. S. Nicholson and I. Shain, Theory of stationary electrode polarography. Single scan and cyclic methods applied to reversible, irreversible, and kinetic systems., *Anal. Chem.* **36**, 706–23, (1964).

[3] J. Crank, *The Mathematics of Diffusion.* (Oxford University Press, Oxford, 1975).

[4] A. Fick, On liquid diffusion, *Lond. Edinb. Dubl. Phil. Mag.* **10**, 30–39, (1855).

[5] A. Fick, Ueber diffusion, *Ann. Phys. (Berlin).* **94**, 59–86, (1855).

[6] A. Einstein, Motion of suspended particles in stationary liquids required from the molecular kinetic theory of heat., *Ann. Phys.* **17**, 549–560, (1905).

Chapter 3

Numerical Solution of the Model System

The goal of the electrochemical modelling in this chapter is to solve the mathematical model developed in the previous chapter in order to obtain the form of the algebraic (containing no derivatives) function $C(X, T)$, i.e., to determine how the concentration of the chemical species varies in space and in time. From this, other information, such as the current passed at the electrode, can be inferred. A number of analytical techniques exist that may be used for solving partial differential equations (PDEs) of the type encountered in electrochemical problems, including integral transform methods such as the Laplace transform, and the method of separation of variables. Unfortunately these techniques are not applicable in all cases and so it is often necessary to resort to the use of numerical methods to find a solution.

Numerical methods are a family of mathematical techniques for solving complicated problems approximately by the repetition of the elementary mathematical operations $(+, -, \times, \div)$. In the past, numerical solution of a PDE was extremely time consuming, requiring many man-hours of tedious calculation. However the utility of numerical methods has greatly increased since the advent of programmable computers. In the field of electrochemistry, two main techniques are used for simulation purposes: the finite difference method and the finite element method, though the former is by far the more popular and will be used exclusively throughout this book.

3.1. Finite Differences

Under a finite difference scheme, a continuous function is represented as a series of finite steps. In our case, the function we are interested in is the concentration as it varies in space, $C(X)$, and its first and second derivatives; we shall ignore for the moment the fact that C also varies in time and will treat the derivatives of C as total rather than partial. To approximate the function $C(X)$, the one-dimensional space under consideration is divided into a set n of discrete points, $X_0, X_1, \ldots, X_{n-1}$, each separated by a constant distance, ΔX, such that for any point, i

$$X_{i+1} - X_i = X_i - X_{i-1} = \Delta X \tag{3.1}$$

and where

$$n = 1 + \frac{X_{\max} - X_0}{\Delta X} \tag{3.2}$$

Under this scheme, the point X_0 corresponds to the electrode surface, and the point X_{n-1} corresponds to the outer spatial boundary. We call this set of points the *spatial grid*. One important consequence of this process of discretisation is that the function $C(X)$ may be easily represented in a computer's memory as a simple array (or similar container) of floating point values of size n. The points are labelled as $X_0, \ldots, X_{n-1}$ rather than as $X_1, \ldots, X_n$ since arrays are zero-indexed (begin with element 0 rather than 1) in C++ and related languages. Figure 3.1 illustrates the discretisation of a function. It is implicit in this scheme that the value of $C(X)$ varies linearly between adjacent points.

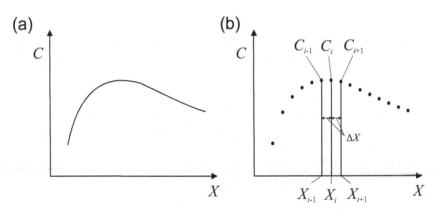

Fig. 3.1. (a) Continuously varying concentration as a function of space, $C(X)$; (b) discrete values of concentration in a discretised space with constant spacing ΔX. Note that in the discrete case, the concentration is assumed to vary linearly between adjacent points.

Let us now consider the first derivative; the mathematical definition of the derivative is the ratio of the change in concentration to the change of distance, i.e.,

$$\left(\frac{dC}{dX}\right)_{X_i} = \lim_{\Delta X \to 0} \frac{C(X_{i+1}) - C(X_i)}{\Delta X} \tag{3.3}$$

From this definition, we can approximate the first derivative as

$$\left(\frac{dC}{dX}\right)_{X_i} \approx \frac{C_{i+1} - C_i}{\Delta X} \tag{3.4}$$

where for simplicity of notation, we express the concentration at point i, $C(X_i)$, as C_i. This is called the *forward difference approximation* of the first derivative. Alternatively we could approximate the derivative as

$$\left(\frac{dC}{dX}\right)_{X_i} \approx \frac{C_i - C_{i-1}}{\Delta X} \tag{3.5}$$

which is called the *backward difference approximation* and is equally valid.

We can define the *truncation* error, Γ, implicit in making these approximations (considering the forward difference scheme as an example) as

$$\Gamma = \left(\frac{dC}{dX}\right)_{X_i} - \frac{C_{i+1} - C_i}{\Delta X} \tag{3.6}$$

We can examine the nature of this error by considering the Taylor series expansion of the concentration:

$$C_{i+1} = C_i + \Delta X \left(\frac{dC}{dX}\right)_{X_i} + \frac{(\Delta X)^2}{2}\left(\frac{d^2C}{dX^2}\right)_{X_i} + \frac{(\Delta X)^3}{6}\left(\frac{d^3C}{dX^3}\right)_{X_i} + \cdots \tag{3.7}$$

Rearranging it, we get

$$\left(\frac{dC}{dX}\right)_{X_i} - \frac{C_{i+1} - C_i}{\Delta X} = -\frac{\Delta X}{2}\left(\frac{d^2C}{dX^2}\right)_{X_i} - \frac{(\Delta X)^2}{6}\left(\frac{d^3C}{dX^3}\right)_{X_i} - \cdots \tag{3.8}$$

From Eq. (3.6), the error is therefore equal to

$$\Gamma_{\text{forward}} = -\frac{\Delta X}{2}\left(\frac{d^2C}{dX^2}\right)_{X_i} - \frac{(\Delta X)^2}{6}\left(\frac{d^3C}{dX^3}\right)_{X_i} - \cdots \tag{3.9}$$

Using the same procedure for the backward difference scheme yields

$$C_{i-1} = C_i - \Delta X \left(\frac{dC}{dX}\right)_{X_i} + \frac{(\Delta X)^2}{2}\left(\frac{d^2C}{dX^2}\right)_{X_i} - \frac{(\Delta X)^3}{6}\left(\frac{d^3C}{dX^3}\right)_{X_i} + \cdots \tag{3.10}$$

$$\Gamma_{backward} = \frac{\Delta X}{2} \left(\frac{d^2 C}{dX^2} \right)_{X_i} - \frac{(\Delta X)^2}{6} \left(\frac{d^3 C}{dX^3} \right)_{X_i} + \cdots \tag{3.11}$$

Now since in general, $\Delta X \ll 1$, it is necessarily true that $\Delta X > (\Delta X)^2 > (\Delta X)^3$, so in each of these series the first term (the term in ΔX) is by far the largest source of error. Therefore for both the forward and backward difference schemes we may write

$$\Gamma \propto \Delta X \tag{3.12}$$

Consequently, the smaller the spatial increment, ΔX, the smaller the error, and therefore the more closely the approximation will agree with the true value of $\left(\frac{dC}{dX} \right)_{X_i}$. This analysis agrees with the definition of the derivative given in Eq. (3.3).

3.1.1. *Central differencing*

An alternative differencing scheme may be arrived at by considering the average derivative given by the forward and backward schemes; this is called the *central difference approximation*:

$$\left(\frac{dC}{dX} \right)_{central} = \frac{1}{2} \left[\left(\frac{dC}{dX} \right)_{forward} + \left(\frac{dC}{dX} \right)_{backward} \right] \tag{3.13}$$

So

$$\left(\frac{dC}{dX} \right)_{central} = \frac{C_{i+1} - C_{i-1}}{2\Delta X} + \Gamma_{central} \tag{3.14}$$

The error is then the average of the errors of both schemes:

$$\Gamma_{central} = \frac{\Gamma_{forward} + \Gamma_{backward}}{2} \tag{3.15}$$

It can be seen that the first (ΔX) terms in Eqs. (3.9) and (3.11) exactly cancel (as do all the terms in odd powers of ΔX), leaving a total error:

$$\Gamma_{central} = -\frac{(\Delta X)^2}{6} \left(\frac{d^3 C}{dX^3} \right)_{X_i} - \frac{(\Delta X)^4}{120} \left(\frac{d^5 C}{dX^5} \right)_{X_i} + \cdots \tag{3.16}$$

Again the leading term is the most significant, so for the central difference scheme,

$$\Gamma \propto (\Delta X)^2 \tag{3.17}$$

This is a second-order error, which is better than the first-order error given by either the forward or backward schemes, and so the central scheme provides a more accurate approximation of the true value of the first derivative. Throughout this book we shall therefore use the approximation

$$\left(\frac{dC}{dX}\right)_{X_i} \approx \frac{C_{i+1} - C_{i-1}}{2\Delta X} \tag{3.18}$$

3.1.2. *Second derivative*

The second derivative of the concentration can be approximated by summing the Taylor series of the forward and backward schemes (Eqs. (3.7) and (3.10)):

$$C_{i+1} + C_{i-1} = 2C_i + (\Delta X)^2 \left(\frac{d^2C}{dX^2}\right)_{X_i} + \frac{(\Delta X)^4}{12} \left(\frac{d^4C}{dX^4}\right)_{X_i} + \cdots \tag{3.19}$$

So by rearranging it, we get

$$\left(\frac{d^2C}{dX^2}\right)_{X_i} = \frac{C_{i-1} - 2C_i + C_{i+1}}{(\Delta X)^2} + \Gamma \tag{3.20}$$

where here

$$\Gamma = -\frac{(\Delta X)^2}{12} \left(\frac{d^4C}{dX^4}\right)_{X_i} + \cdots \tag{3.21}$$

so the second derivative of concentration may be approximated as

$$\left(\frac{d^2C}{dX^2}\right)_{X_i} \approx \frac{C_{i-1} - 2C_i + C_{i+1}}{(\Delta X)^2} \tag{3.22}$$

again with error $\Gamma \propto (\Delta X)^2$.

The approximations that we have developed here for the first and second derivatives at point i consider only three points: i itself, and the two adjacent points. It is of course possible to formulate finite difference equations over a greater number of points than this using the same general technique, and doing so usually results in a higher-order error (a more accurate approximation). However, assuming that ΔX is sufficiently small, the increase in accuracy is minimal and may not worth the increase in complexity.

3.2. Time Evolution: Discretising Fick's Second Law

So far in this chapter we have treated the concentration as if it varied with only one independent variable — position, X. As we saw in Chapter 2, modelling of an electrochemical system requires the solution of Fick's second law, a partial differential equation in which C varies as a function of both space, X, and time, T. We must now therefore consider how the system evolves in time. Thus, we are going to study the simulation of the reversible, one-electron reduction of species A at a macroelectrode:

$$A + e^- \rightleftharpoons B \tag{3.23}$$

where the diffusion coefficient of both species has the same value and only species A is initially present. As discussed in the previous chapter, under these conditions only the problem corresponding to species A needs to be solved since the concentration profile of species B can be obtained immediately as $c_B = c_A^* - c_A$.

For the one-dimensional system under consideration, the state of the system (which is represented in a discretised form by the set of concentrations $C_0, \ldots, C_{n-1}$) at any moment in time depends on two things: the spatial boundary conditions (the electrode surface and bulk solution boundaries), and the state of the system at the previous moment in time. We may discretise time in the same manner as space, with a constant interval between adjacent timesteps, ΔT. When modelling, we are interested in the state of the system from the starting time of the experiment, T_0, to its finishing time, $T_{\max}$, so the total number of timesteps, m, is given by

$$m = \frac{T_{\max} - T_0}{\Delta T} \tag{3.24}$$

As previously discussed, at the beginning of the experiment, the system is homogeneous, i.e., the concentration of the starting species A is uniform across the whole space as given by the initial conditions in the previous chapter, $C(X, T = 0) = 1$. Therefore, initially the values of the set of concentrations are all equal to 1. The concentrations at time T_1 are then a function of those at T_0, as given by Fick's second law:

$$\frac{\partial C}{\partial T} = \frac{\partial^2 C}{\partial X^2} \tag{3.25}$$

and subject to the boundary conditions. For the derivative of concentration with respect to time, it is obvious that we must use a backward difference scheme as we have no information about the state of the system at any timesteps beyond the current one. If we let the previous timestep (at which the concentrations are known) be T^{k-1} and the current step at which they are unknown be T^k, then

$$\frac{\partial C_i}{\partial T} \approx \frac{C_i^k - C_i^{k-1}}{\Delta T} \tag{3.26}$$

The discretised version of Fick's second law is then

$$\frac{C_i^k - C_i^{k-1}}{\Delta T} = \frac{C_{i-1} - 2C_i + C_{i+1}}{(\Delta X)^2} \tag{3.27}$$

where the right-hand side comes from Eq. (3.22). There is now a choice as to whether the concentrations on the right-hand side are chosen to be at T^{k-1} or T^k. We will briefly consider the former case, generally known as the *explicit method*, illustrated in Figure 3.2(a), before moving on to the

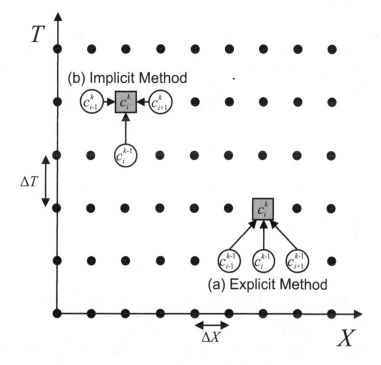

Fig. 3.2. Discrete simulation space-time grid showing concentration terms involved in the equation for C_i^k using the (a) explicit and (b) implicit methods. Grey squares show the term to be solved for and white circles show the terms that appear in the equation.

latter, the *implicit method*, illustrated in Figure 3.2(b), which we will be using throughout the rest of this work.

3.2.1. *The explicit method*

The explicit method consists of solving the equation

$$\frac{C_i^k - C_i^{k-1}}{\Delta T} = \frac{C_{i-1}^{k-1} - 2C_i^{k-1} + C_{i+1}^{k-1}}{(\Delta X)^2} \tag{3.28}$$

which for convenience may be rearranged to

$$C_i^k = C_i^{k-1} + \lambda(C_{i-1}^{k-1} - 2C_i^{k-1} + C_{i+1}^{k-1}) \tag{3.29}$$

where

$$\lambda = \frac{\Delta T}{(\Delta X)^2} \tag{3.30}$$

This is a simple equation to solve, and each value of C_i^k can be solved for independently at every timestep because all of the terms appearing on the right-hand side of the equation are fully known (this is in contrast with the implicit method which we shall come to shortly). Simulation using the explicit method simply consists of iterating from time T_0, where for all i,

$$C_i = 1 \tag{3.31}$$

to timestep $m - 1$, where for $k \geq 0$

$$C_i^k = \begin{cases} 1/(1 + e^{-\theta^{(k)}}) & \text{for } i = 0 \\ C_i^{k-1} + \lambda(C_{i-1}^{k-1} - 2C_i^{k-1} + C_{i+1}^{k-1}) & \text{for } i = 1 \text{ to } n - 2 \\ 1 & \text{for } i = n - 1 \end{cases} \tag{3.32}$$

where the dimensionless applied potential $\theta^{(k)}$ can be different in each timestep (as in cyclic voltammetry) or not (as in chronoamperometry).

While this method is very simple and was more popular before the advent of fast computers, it is only conditionally stable. Numerical stability is guaranteed [1, 2] if

$$\lambda < 0.5 \tag{3.33}$$

but if this condition is not met, the solution will be unstable, diverging from the true solution, and the concentrations will oscillate wildly in both time and space, possibly resulting in negative concentrations, which are obviously physically unrealistic. Consequently, to maintain accuracy, ΔT must decrease with decreasing ΔX, making it inefficient when one considers the small size of ΔX necessary to accurately represent a continuous function [3]. For this reason, the explicit method is of only limited general utility in electrochemistry though it does find use in some specific areas, particularly in the modelling of rotating disc electrodes (see Chapter 8) [4].

3.2.2. *The implicit method*

The *backward implicit* method consists of solving the equation

$$\frac{C_i^k - C_i^{k-1}}{\Delta T} = \frac{C_{i-1}^k - 2C_i^k + C_{i+1}^k}{(\Delta X)^2} \tag{3.34}$$

The only difference from Eq. (3.28) is the timestep at which the concentrations on the right-hand side are taken. Rearranged to a more useful form, it becomes

$$C_i^{k-1} = C_i^k - \lambda(C_{i-1}^k - 2C_i^k + C_{i+1}^k) \tag{3.35}$$

This equation contains a single known value, C_i^{k-1}, and three unknowns, C_{i-1}^k, C_i^k and C_{i-1}^k, and so unlike the explicit method cannot be solved in isolation for any given point, i. At each timestep, k, the complete set of equations for the concentration at every point must be solved simultaneously. The set of equations is

$$
\begin{array}{ccccc}
\vdots & \vdots & \vdots & \vdots & \\
-\lambda C_{43}^k + (1+2\lambda)C_{44}^k - \lambda C_{45}^k & = & C_{44}^{k-1} & \\
-\lambda C_{44}^k + (1+2\lambda)C_{45}^k - \lambda C_{46}^k & = & C_{45}^{k-1} & (3.36) \\
-\lambda C_{45}^k + (1+2\lambda)C_{46}^k - \lambda C_{47}^k & = & C_{46}^{k-1} & \\
\vdots & \vdots & \vdots & \vdots &
\end{array}
$$

with boundary conditions

$$C_0^{(k)} = 1/(1 + e^{-\theta^{(k)}})$$

$$C_{n-1} = 1$$

We therefore have n unknowns and n non-degenerate equations and are thus able to solve simultaneously for the whole set of concentrations. While more complicated than the explicit scheme, the implicit scheme is not subject to the constraint imposed by Eq. (3.33) and is therefore unconditionally stable, which makes it much more computationally efficient.

3.3. The Thomas Algorithm

The standard method of solving a set of simultaneous equations with a computer is to cast the set as a matrix equation. For example, the set

$$6x_1 + 3x_2 + 4x_3 = 67$$
$$9x_1 + 4x_2 + 1x_3 = 25$$
$$7x_1 + 8x_2 + 7x_3 = 78$$

will be cast as the equation

$$\begin{pmatrix} 6 & 3 & 4 \\ 9 & 4 & 1 \\ 7 & 8 & 7 \end{pmatrix} \begin{pmatrix} x_1 \\ x_2 \\ x_3 \end{pmatrix} = \begin{pmatrix} 67 \\ 25 \\ 78 \end{pmatrix} \tag{3.37}$$

which is an equation of the form

$$\mathbf{Ax} = \mathbf{b} \tag{3.38}$$

To find the set of solutions, $\mathbf{x}$, we pre-multiply both sides of the equation by $\mathbf{A}^{-1}$, the inverse of $\mathbf{A}$:

$$\mathbf{x} = \mathbf{A}^{-1}\mathbf{b} \tag{3.39}$$

For this simple example with only three unknowns, we could calculate the product $\mathbf{A}^{-1}\mathbf{b}$ by hand; however, for systems with a large number of unknowns, this is not practical. For a dense matrix $\mathbf{A}$, one with mostly non-zero values, the normal way to calculate $\mathbf{A}^{-1}\mathbf{b}$ is the general *Gaussian elimination algorithm* [3]. This algorithm takes a number of mathematical operations that is proportional to n^3 where n is the number of unknowns.

The equation set we consider here (3.36) has n equations so $\mathbf{A}$ will be an $n \times n$ matrix. As n may be arbitrarily large, the cost of calculating $\mathbf{A}^{-1}$ by Gaussian elimination may be very high — it will take a long time to compute. Fortunately, the equation set has a very useful property of which we can take advantage. For each spatial point i, the equation is of the form

$$\alpha_i C_{i-1}^k + \beta_i C_i^k + \gamma_i C_{i+1}^k = \delta_i^k \qquad (3.40)$$

where, in our problem, for $i = 1$ to $n - 2$,

$$\alpha_i = -\lambda, \qquad \beta_i = 1 + 2\lambda, \qquad \gamma_i = -\lambda, \qquad \delta_i^k = C_i^{k-1} \qquad (3.41)$$

So α, β and γ do not vary with i, except at the electrode surface boundary,

$$\alpha_0 = 0, \qquad \beta_0 = 1, \qquad \gamma_0 = 0, \qquad \delta_0^{(k)} = \frac{1}{1 + e^{-\theta(k)}} \qquad (3.42)$$

and outer spatial boundary

$$\alpha_{n-1} = 0, \qquad \beta_{n-1} = 1, \qquad \gamma_{n-1} = 0, \qquad \delta_{n-1} = 1 \qquad (3.43)$$

Note that in the following chapter, we will utilise spatial grids with unequal intervals such that ΔX varies with i; in that case, α, β and γ will also vary with i. In the equation for C_i^k, the coefficients for the terms $C_0^{(k)}$ to C_{i-2}^k and C_{i+2}^k to C_{n-1} are all 0; this makes $\mathbf{A}$ a *sparse matrix* where only the three central diagonals contain non-zero elements. The equation set may be written as

$$
\begin{pmatrix}
1 & 0 & 0 & 0 & 0 & 0 & 0 & 0 & 0 & 0 \ldots \\
\alpha & \beta & \gamma & 0 & 0 & 0 & 0 & 0 & 0 & 0 \ldots \\
0 & \alpha & \beta & \gamma & 0 & 0 & 0 & 0 & 0 & 0 \ldots \\
& & \ddots & \ddots & \ddots & & & & & \\
\ldots 0 & 0 & \alpha & \beta & \gamma & 0 & 0 & 0 & 0 \ldots \\
\ldots 0 & 0 & 0 & \alpha & \beta & \gamma & 0 & 0 & 0 \ldots \\
\ldots 0 & 0 & 0 & 0 & \alpha & \beta & \gamma & 0 & 0 \ldots \\
& & & & \ddots & \ddots & \ddots & & & \\
\ldots 0 & 0 & 0 & 0 & 0 & 0 & \alpha & \beta & \gamma & 0 \\
\ldots 0 & 0 & 0 & 0 & 0 & 0 & 0 & \alpha & \beta & \gamma \\
\ldots 0 & 0 & 0 & 0 & 0 & 0 & 0 & 0 & 0 & 1
\end{pmatrix}
\begin{pmatrix}
C_0^{(k)} \\
C_1^k \\
C_2^k \\
\vdots \\
C_{44}^k \\
C_{45}^k \\
C_{46}^k \\
\vdots \\
C_{n-3}^k \\
C_{n-2}^k \\
C_{n-1}
\end{pmatrix}
=
\begin{pmatrix}
1/(1 + e^{-\theta(k)}) \\
C_1^{k-1} \\
C_2^{k-1} \\
\vdots \\
C_{44}^{k-1} \\
C_{45}^{k-1} \\
C_{46}^{k-1} \\
\vdots \\
C_{n-3}^{k-1} \\
C_{n-2}^{k-1} \\
1
\end{pmatrix}
$$

$$(3.44)$$

Equations of this type are called *tridiagonal matrix equations* and may be solved in a more efficient manner using the *Thomas algorithm* [3] which is a simplified version of Gaussian elimination.

Rather than directly calculating the product $\mathbf{A}^{-1}\mathbf{b}$, we use a process called LU factorisation, to factor $\mathbf{A}$ as the product of two other matrices: a lower and an upper triangular matrix, so that $\mathbf{A} = \mathbf{LU}$:

$$
\begin{pmatrix}
\beta_0 & \gamma_0 & & & \\
\alpha_1 & \beta_1 & \gamma_1 & & \\
& \ddots & \ddots & \ddots & \\
& & \alpha_{n-2} & \beta_{n-2} & \gamma_{n-2} \\
& & & \alpha_{n-1} & \beta_{n-1}
\end{pmatrix}
$$

$$
=
\begin{pmatrix}
\beta_0' & & & & \\
\alpha_1 & \beta_1' & & & \\
& \ddots & \ddots & & \\
& & \alpha_{n-2} & \beta_{n-2}' & \\
& & & \alpha_{n-1} & \beta_{n-1}'
\end{pmatrix}
\begin{pmatrix}
1 & \gamma_0' & & & \\
& 1 & \gamma_1' & & \\
& & \ddots & \ddots & \\
& & & 1 & \gamma_{n-2}' \\
& & & & 1
\end{pmatrix}
\tag{3.45}
$$

where β_i' and γ_i' are modified coefficients that have different values to the β_i and γ_i coefficients respectively. This process leaves us with the equation $\mathbf{LUx} = \mathbf{b}$. The algorithm then consists of two steps. First, we pre-multiply both sides of the equation by $\mathbf{L}^{-1}$:

$$
\mathbf{Ux} = \mathbf{L}^{-1}\mathbf{b}
\tag{3.46}
$$

Then we calculate the product $\mathbf{d} = \mathbf{L}^{-1}\mathbf{b}$. The equation $\mathbf{Ux} = \mathbf{d}$ may be written as

$$
\begin{pmatrix}
1 & \gamma_0' & & & \\
& 1 & \gamma_1' & & \\
& & \ddots & \ddots & \\
& & & 1 & \gamma_{n-2}' \\
& & & & 1
\end{pmatrix}
\begin{pmatrix}
C_0^{(k)} \\
C_1^{k} \\
\vdots \\
C_{n-2}^{k} \\
C_{n-1}'
\end{pmatrix}
=
\begin{pmatrix}
\delta_0'^{(k)} \\
\delta_1'^{k} \\
\vdots \\
\delta_{n-2}'^{k} \\
\delta_{n-1}'
\end{pmatrix}
\tag{3.47}
$$

where the vector $\mathbf{d}$ consists of the set of values $\delta_i'^{k}$, which are different to the set of values δ_i^{k}. To complete this first step then, we must calculate these modified γ_i' and $\delta_i'^{k}$ values. The γ_i' values in the $\mathbf{U}$ matrix may be calculated simply from

$$
\gamma_i' = \begin{cases}
\gamma_i/\beta_i & \text{for } i = 0 \\
\gamma_i/(\beta_i - \gamma_{i-1}'\alpha_i) & \text{for } i = 1, 2, \ldots, n-2
\end{cases}
\tag{3.48}
$$

The vector $\mathbf{d}$ (the set of $\delta_i'^k$ values) could be calculated by first evaluating the lower diagonal matrix, $\mathbf{L}$, inversing it, then post-multiplying the result by $\mathbf{b}$. This would be computationally expensive by normal means, but fortunately, the $\delta_i'^k$ values may be calculated directly from

$$\delta_i'^k = \begin{cases} \delta_i^{(k)}/\beta_i & \text{for } i = 0 \\ (\delta_i^k - \delta_{i-1}'^k \alpha_i)/(\beta_i - \gamma_{i-1}' \alpha_i) & \text{for } i = 1, 2, \ldots, n-1 \end{cases} \tag{3.49}$$

This first step is called the *forward sweep*. Note that for the specific model studied in this chapter, the surface boundary condition ($\beta_0 = 1$, $\gamma_0 = 0$) means that for $i = 0$, the modified coefficients will have the values $\gamma_0' = 0$, $\delta_0'^{(k)} = \delta_0^{(k)}$. Likewise, as a consequence of the outer spatial boundary condition, $\delta_{n-1}' = \delta_{n-1}$. In the second step of the algorithm, we pre-multiply both sides of Eq. (3.46) by $\mathbf{U}^{-1}$:

$$\mathbf{x} = \mathbf{U}^{-1}\mathbf{d} \tag{3.50}$$

And so we calculate the product $\mathbf{U}^{-1}\mathbf{d}$ to find the vector $\mathbf{x}$ of unknown C_i^k values. These are calculated simply as

$$C_i^k = \begin{cases} \delta_{n-1}' & \text{for } i = n-1 \\ \delta_i'^k - \gamma_i' C_{i+1}^k & \text{for } i = n-2, n-3, \ldots, 0 \end{cases} \tag{3.51}$$

This second step is called *back substitution*.

In general, when implementing this algorithm, rather than storing several $n \times n$ matrices in the computer's memory, we simply store the α_i, β_i and γ_i in three $1 \times n$ containers (see Chapter 4). At each successive timestep, the values of these coefficients are calculated, and then the γ_i coefficients are modified in place according to Eq. (3.48) (i.e., the γ_i' values overwrite the γ_i values — this removes the need for an additional container to store the γ_i' values separately). Further $1 \times n$ containers can store the $\delta_i^{(k)}$ and $\delta_i'^{(k)}$ values. At the start of each timestep, the former container holds the set of unmodified $\delta_i^{(k)}$ values $[1/(1 + e^{-\theta^{(k)}}), C_1^{k-1}, \ldots, c_{n-2}^{k-1}, 1]$. These are then modified according to Eq. (3.49). The desired solution (the concentration profile at the current timestep, that is, the C_i^k values) ends up in a container that originally held the concentration profile from the previous timestep. These are in turn used as the set of $\delta_i^{(k)}$ values in the next timestep (except for the changes at $i = 0$ and $i = n-1$ to account for the boundary conditions).

For the particular example simulation that we study in this chapter, α_i, β_i and γ_i do not vary with i. We need only a $1 \times n$ container which will store the modified γ_i' coefficients. Further, these coefficients are invariant with time so the set of γ_i' coefficients need only to be calculated once at the start of a simulation and not at every timestep.

3.4. Simulation Procedure

To perform a simulation of a cyclic voltammetry experiment, we solve Eq. (3.44) repeatedly, updating the value of the potential, θ^k, at each subsequent timestep. In order to do this, five parameters must be specified: the initial potential, θ_i, vertex potential, θ_v, scan rate, σ, and the space and time increments, ΔX and ΔT respectively. We may describe the first three as being experimental parameters, as changing them corresponds to a change in the corresponding experimental system. The last two parameters may be described as accuracy parameters, as they control how accurate the simulation is. From the experimental parameters we can determine the total dimensionless time that the simulated cyclic voltammetry experiment lasts for:

$$T_{\max} = \frac{2|\theta_v - \theta_i|}{\sigma} \tag{3.52}$$

The factor of 2 comes from the fact that the potential sweeps from θ_i to θ_v and back again at a constant rate σ. The total number of timesteps is therefore

$$m = \frac{T_{\max}}{\Delta T} = \frac{2|\theta_v - \theta_i|}{\sigma \Delta T} \tag{3.53}$$

We could explicitly specify ΔT; however, if the scan rate is increased, then there would be a decrease in the total number of timesteps, which would decrease the accuracy of the simulation. A superior alternative is to specify the change in potential per timestep, $\Delta \theta$, which results in a fixed number of timesteps for the simulation and therefore a fixed accuracy, independent of the value of σ. The value of ΔT is then

$$\Delta T = \frac{\Delta \theta}{\sigma} \tag{3.54}$$

In Section 3.5 we will investigate how to choose reasonable values of both $\Delta \theta$ and ΔX. Suitable values for the initial and vertex potential in the study of an electro-reduction process are $\theta_i = 20$ and $\theta_v = -20$, which roughly correspond to real values of $E_i - E_f^0 = 0.5$ V and $E_v - E_f^0 = -0.5$ V at 298 K.

At each timestep k, the potential, θ^k, is given by

$$\theta^k = \begin{cases} \theta_i & \text{for } k = 0 \\ \theta^{k-1} - \Delta\theta & \text{for } 0 < k \le m/2 \\ \theta^{k-1} + \Delta\theta & \text{for } m/2 < k \le m \end{cases} \quad (3.55)$$

The primary output of the simulation is the dimensionless current, J; plotting J against θ produces a cyclic voltammogram. J is proportional to the concentration gradient at the surface:

$$J = J^k = -\left(\frac{C_1^k - C_0^k}{\Delta X}\right) \quad (3.56)$$

This is a low-order (two-point) approximation with error proportional to ΔX, but it is quite accurate as long as ΔX is sufficiently small. As an alternative, an asymmetrical three-point discretisation may be used [5]:

$$J = -\left(\frac{-C_2^k + 4C_1^k - 3C_0^k}{2\Delta X}\right) \quad (3.57)$$

which has error proportional to $(\Delta X)^2$. It has been established [6] that the use of higher-order expressions to calculate the flux does not necessarily lead to greater accuracy, since the limiting factor is the order of the finite difference approximations of the spatial derivatives used to represent Fick's second law. As we have used second-order finite difference approximations throughout, a second-order current approximation is the most reasonable choice. In any case, since the determination of the current is a simple calculation performed after, and independently of, the more complicated numerical procedures used at each timestep, any order of approximation for the current may be used without introducing additional complexity to the simulation.

3.4.1. *Example program*

The following is a simple C++ program to simulate the cyclic voltammetry of a one-electron reversible reduction at a macroelectrode where the redox species have equal diffusion coefficients. The program starts by specifying the values of the five input parameters (θ_i, θ_v, σ, ΔX and $\Delta\theta$) and calculating the values of ΔT (the size of the timestep), and of n and m (the number of space- and time-steps respectively). Next, the γ_i' coefficients are calculated and stored in a container and three further containers are created to store the δ_i and δ_i' coefficients and concentrations (this is done for the sake of clarity since a single container would serve by overwriting the values appropriately); additionally, a text file is created to store the flux that is calculated at the end of each timestep. With all this taken care of, the simulation proper begins. The simulation consists of a loop that runs once for each timestep. Each loop iteration consists of four steps: incrementing the potential; calculating the $\delta_i'^{(k)}$ values in the forward sweep; calculating the concentration profile for this timestep in the back substitution; and finally calculating the flux and outputting it to a text file.

Once the program has run, which could take several seconds depending on the inputs, the result will be a text file consisting of a series of potential-flux data points. These data can be plotted in any spreadsheet or graphing program to yield a cyclic voltammogram. A brief introduction to C++ may be found in Appendix A.

```cpp
#include <fstream>    // Header for file output
#include <vector>     // Header for vectors
#include <cmath>      // Header for sqrt(), fabs()

int main()
{
    // Specify simulation parameters
    double theta_i = 20.0;
    double theta_v = -20.0;
    double sigma = 100.0;
    double deltaX = 2e-4;
    double deltaTheta = 0.02;

    //Calculate other parameters
    double deltaT = deltaTheta / sigma;
    double maxT = 2.0 * fabs(theta_v - theta_i) / sigma;
    double maxX = 6.0*sqrt(maxT);
    int n = (int)( 1.0 + maxX / deltaX );  // number of spacesteps
    int m = (int)( maxT / deltaT );  // number of timesteps

    // Calculate Thomas coefficients
    double lambda = deltaT / (deltaX*deltaX);
    double alpha = -lambda;
    double beta = 2.0*lambda + 1.0;
    double gamma = -lambda;
```

```cpp
// Create containers
std::vector<double> g_mod(n-2, 0.0);
std::vector<double> delta(n-1, 1.0);
std::vector<double> d_mod(n-1, 0.0);
std::vector<double> C(n, 1.0); // concentration profile

// Modify gamma coefficients
g_mod[0] = 0; // boundary condition
for(int i=1; i<n-2; i++) {
    g_mod[i] = gamma / (beta - g_mod[i-1] * alpha);
}


// Open file to output CV
std::ofstream CV("CV_Output.txt");

// BEGIN SIMULATION
double Theta = theta_i;

for(int k=0; k<m; k++)
{
    if( k < m / 2 ) { Theta -= deltaTheta; }
    else            { Theta += deltaTheta; }

    // Forward sweep - create modified deltas
    d_mod[0] = 1.0 / (1.0 + exp(-Theta));
    for(int i=1; i<n-1; i++) {
        d_mod[i] = ( delta[i] - d_mod[i-1]*alpha )
            / ( beta - g_mod[i-1] * alpha );
    }

    // Back Substitution
    C[n-1] = 1.0;
    for(int i=n-2; i>=0; i--) {
        C[i] = d_mod[i] - g_mod[i] * C[i+1];

        delta[i] = C[i];
    }

    //Output current
    double flux = -(-C[2] + 4.0*C[1] -3.0*C[0]) / (2.0*deltaX);
    CV << Theta << "\t" << flux << "\n";
}
// END SIMULATION
}
```

3.5. Testing Simulations

The approximation of derivatives by finite differences will inherently involve some error in the simulation results with respect to the exact solution of the problem. By decreasing the intervals of the spatial and time grids, discretisation errors will decrease and the simulation results will approach the exact solution, as much as the rounding errors determined by the machine accuracy allow, at the expense of runtime and memory requirements (as we shall see in Section 3.6).

Errors in the mathematical model, as well as mistakes in the programming process, are also possible. Hence, it is always necessary to test our methods and simulations and to determine optimum *gridding* (ΔX and ΔT) so that a trade-off situation of accuracy, programming effort and runtime is achieved. For this, several indicators and approaches will be discussed in this section.

3.5.1. *General tests*

3.5.1.1. *Self-convergence analysis*

As discussed in Section 3.1, truncation errors decrease as the corresponding time or spatial interval is smaller so that the numerical solution can be expected to approach (converge towards) the exact solution of the problem as ΔX and ΔT are decreased. If so, the numerical method is said to be convergent and the simulation accuracy will increase as these intervals shrink. Conversely, if the size of the intervals is too large, the simulation may produce wildly inaccurate results.

According to the above, a general preliminary test to assess the implementation and performance of our numerical simulations can be based on a *convergence study* where results obtained upon systematic refinement of the spatial and time grids are compared. Thus, for example, Figure 3.3(a) demonstrates how the peak current in cyclic voltammetry, J_p, calculated with the example program given in Subsection 3.4.1, varies with the size of ΔX; an equivalent study can be performed for $\Delta\theta$. As can be seen, the peak current tends to a constant value as the size of the spatial increment decreases. We can comment on whether or not the simulation is converged by examining the rate of change of J_p with respect to ΔX as shown in Figure 3.3(b). We can therefore define the solution to be converged once the value of $dJ_p/d\log(\Delta X)$ falls below some specified threshold.

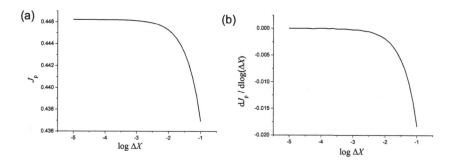

Fig. 3.3. (a) Variation of simulated peak current with size of spatial increment; (b) first derivative of the same. $\sigma = 1$, $\Delta\theta = 0.1$.

3.5.1.2. *Unchanging stoichiometry-weighted total number of particles*

Irrespectively of the features of the chemical and electrochemical reactions and the electrochemical technique, the total number of particles (atoms or molecules) of the participating species, once adequately weighted according to their stoichiometric coefficients s_j, must be constant. This establishes a very general criterion to test the simulations [7]. Thus, for the generic electron transfer reaction

$$s_A A + ne^- \rightleftarrows s_B B \qquad (3.58)$$

it must be fulfilled that

$$\frac{n_A}{s_A} + \frac{n_B}{s_B} = \frac{n_A^*}{s_A} + \frac{n_B^*}{s_B} = const. \qquad (3.59)$$

where n_A and n_B are the total number of moles of species A and B, respectively, at any time of the simulated experiment and n_A^* and n_B^*, their initial total number of moles. In the case of continuous concentrations, the above expression can be rewritten in terms of the species concentration profiles as follows

$$\frac{1}{s_A} \int\limits_{V_{total}} c_A(q,t)\, dV + \frac{1}{s_B} \int\limits_{V_{total}} c_B(q,t)\, dV = \frac{n_A^*}{s_A} + \frac{n_B^*}{s_B} = const. \qquad (3.60)$$

where q refers to all the spatial coordinates and V_{total} to the total volume of the solution.

Condition (3.60) in the one-dimension, equally-spaced simulation space from 0 to X_{max} introduced in Section 3.1 in terms of the dimensionless distance and concentrations (Chapter 2) and using the trapezium rule to approximate the integration (see Chapter 9) leads to that for each timestep k

$$\frac{1}{s_A}\frac{\Delta X}{2}\sum_{i=0}^{n-2}\left(C_{i,A}^k + C_{i+1,A}^k\right) + \frac{1}{s_B}\frac{\Delta X}{2}\sum_{i=0}^{n-2}\left(C_{i,B}^k + C_{i+1,B}^k\right)$$

$$= \frac{1}{s_A}X_{\max} + \frac{1}{s_B}\frac{c_B^*}{c_A^*}X_{\max} \tag{3.61}$$

so that one can define the following reference quantity

$$Q_n^k = \frac{\Delta X}{2}\sum_{i=0}^{n-2}\left[\frac{1}{s_A}\left(C_{i,A}^k + C_{i+1,A}^k\right) + \frac{1}{s_B}\left(C_{i,B}^k + C_{i+1,B}^k\right)\right]$$

$$- \left(\frac{1}{s_A}X_{\max} + \frac{1}{s_B}\frac{c_B^*}{c_A^*}X_{\max}\right) \tag{3.62}$$

the Q_n^k-value approaching zero as the numerical solution is improved.

Note that the above indicator can be extended to electrode reactions complicated by homogeneous chemical reactions that alter the chemical entity of the redox species. Thus, for example, for the generic first-order EC mechanism

$$s_A A + ne^- \rightleftarrows s_B B$$
$$s_B B \rightleftarrows s_Y Y \tag{3.63}$$

the reference quantity will read as

$$Q_n^k = \frac{\Delta X}{2}\sum_{i=0}^{n-2}\left[\frac{1}{s_A}\left(C_{i,A}^k + C_{i+1,A}^k\right) + \frac{1}{s_B}\left(C_{i,B}^k + C_{i+1,B}^k\right) + \frac{1}{s_Y}\left(C_{i,Y}^k + C_{i+1,Y}^k\right)\right]$$

$$- \left(\frac{1}{s_A}X_{\max} + \frac{1}{s_B}\frac{c_B^*}{c_A^*}X_{\max} + \frac{1}{s_Y}\frac{c_Y^*}{c_A^{*}}X_{\max}\right) \tag{3.64}$$

and in general for any ET mechanism with j participating species (that is, the redox species and those they transform into via chemical reactions) as

$$Q_n^k = \frac{\Delta X}{2}\sum_{j \text{ species}}\frac{1}{s_j}\sum_{i=0}^{n-2}\left(C_{i,j}^k + C_{i+1,j}^k\right) - X_{\max}\sum_{j \text{ species}}\frac{1}{s_j}\frac{c_j^*}{c_A^*} \tag{3.65}$$

Alternatively (or additionally), one can monitor the magnitude of the oscillations of the weighted total number of moles along the simulation through the following quantity

$$Q_n = \frac{\max\left(\tilde{n}^k\right) - \min\left(\tilde{n}^k\right)}{\operatorname{avg}\left(\tilde{n}^k\right)} \tag{3.66}$$

where

$$\tilde{n}^k = \frac{\Delta X}{2} \sum_{j\ \text{species}} \frac{1}{s_j} \sum_{i=0}^{n-2} \left(C_{i,j}^k + C_{i+1,j}^k\right) \tag{3.67}$$

and $\max\left(\tilde{n}^k\right)$, $\min\left(\tilde{n}^k\right)$ and $\operatorname{avg}\left(\tilde{n}^k\right)$ refer to the maximum, minimum and average values of $\tilde{n}^k$, respectively. Hence, the Q_n-value will decrease as the simulation is improved.

3.5.1.3. *Unchanging stoichiometry-weighted total concentration*

Provided that the diffusion coefficient of the participating species are *equal* and that any (electro)chemical reaction conserves the number of species particles in solution (no adsorption, aggregation, dissolution,...) [8, 9], it can be demonstrated that the total concentration of species remains uniform and constant in diffusion-only and diffusion-convection systems. When migration takes place, this is also true as long as the electroneutrality condition holds [9, 10]. Thus, for a generic electrode reaction with j participating species and $s_j = 1$, it is fulfilled that

$$\sum_{j\ \text{species}} c_j(q, t) = \sum_{j\ \text{species}} c_j^* = const. \tag{3.68}$$

that enables us to define the following reference quantity for each point i of the spatial grid

$$Q_{c,i}^k = \sum_{j\ \text{species}} \left(C_{i,j}^k - \frac{c_j^*}{c_A^*}\right) \tag{3.69}$$

the value of which will approach zero as the simulation is more accurate. Extending the above criterion to the more general electrode reaction (3.58) with non-unity stoichiometries, one obtains that [7]

$$\frac{c_A(q, t)}{s_A} + \frac{c_B(q, t)}{s_B} = \frac{c_A^*}{s_A} + \frac{c_B^*}{s_B} = const. \tag{3.70}$$

so that

$$Q_{c,i}^k = \frac{1}{s_A}\left(C_{i,A}^k - 1\right) + \frac{1}{s_B}\left(C_{i,B}^k - \frac{c_B^*}{c_A^*}\right) \tag{3.71}$$

Note that relationships (3.68) and (3.70) can be used to check the simulated concentration profiles or, also interestingly, to reduce the number of variables by one (see Subsection 2.2.1), making the coding and simulation simpler, more rapid and less demanding [9, 11].

3.5.2. *Specific tests: Analytical solutions and reference values*

In many cases, some property of the output of the simulation is described by a known analytical or empirically-derived expression, at least for particular or limit situations of the system under study. These can be used to test if the simulation output is correct through a convergence study. Note that this also enables us to test the *correctness* of the simulation since the programmer may have made a mistake and the simulation may be modelling something other than what was intended. In the following, major references are reviewed for the case of a mono-electronic, reversible electron transfer $(A + e^- \rightleftarrows B)$ where redox species are only subject to linear diffusion, the case considered in this chapter. For other conditions involving electrode kinetics (Chapter 4), convergent spherical diffusion (Chapter 4), steady state conditions (Chapter 4), two-dimensional systems (Chapter 9), coupled chemical reactions (Chapters 5 and 6), migrational mass transport (Chapter 7) or controlled convection (Chapter 8), reference expressions and values will be given in the corresponding chapter.

3.5.2.1. *Any electrode kinetics*

In chronoamperometry under limiting current conditions $(\theta \to -\infty$ for electro-reductions), the dimensionless current-time response $(J(T))$ and concentration profile of the redox species $(C_A(X,T)$ and $C_B(X,T))$ are given by

$$|J(T)| = \frac{1}{\sqrt{\pi T}} \tag{3.72}$$

$$C_A(X,T) = \text{erf}\left(\frac{X}{2\sqrt{T}}\right) \tag{3.73}$$

$$C_B(X,T) = \frac{c_B^*}{c_A^*} + \frac{1}{\sqrt{d_B}}\text{erfc}\left(\frac{X}{2\sqrt{d_B T}}\right) \tag{3.74}$$

with $\mathrm{erf}(x)$ being the error function and $\mathrm{erfc}(x)$ the complementary error function: $\mathrm{erfc}\,(x) = 1 - \mathrm{erf}\,(x) = 1 - \frac{2}{\sqrt{\pi}} \int_0^x e^{-t^2} dt$. Thus, the error of the simulated output can be monitored through the quantities

$$Q_{\mathrm{J}}^k = \frac{J_{\mathrm{sim}}^k - J(k\Delta T)}{J(k\Delta T)} \times 100 \qquad (3.75)$$

$$Q_{\mathrm{C,i}}^k = \frac{C_{\mathrm{i,sim}}^k - C(i\Delta X, k\Delta T)}{C(i\Delta X, k\Delta T)} \times 100 \qquad (3.76)$$

where the subscript *sim* refers to the simulated values.

It is worth highlighting that chronoamperometry under limiting current conditions is a demanding situation to test our simulations given that the concentration profile changes dramatically just after the application of the potential step: the surface concentration of the reactant species A varies from the bulk value c_A^* to *ca.* 0 in a single timestep.

3.5.2.2. *Fully reversible electron transfers*

The dimensionless chronoamperometric response of a reversible electron transfer at a macroelectrode for any value of the potential pulse applied is given by

$$J(T) = \frac{C_{\mathrm{A}}(0) - 1}{\sqrt{\pi T}} \qquad (3.77)$$

where $C_{\mathrm{A}}(0)$ is the dimensionless surface concentration

$$C_{\mathrm{A}}(0) = \frac{e^\theta \left(1 + \sqrt{d_B} c_B^*/c_A^*\right)}{\sqrt{d_B} + e^\theta} \qquad (3.78)$$

which fulfills that

$$C_A(0) + \sqrt{d_B}\, C_B(0) = 1 + \sqrt{d_B}\, \frac{c_B^*}{c_A^*} \qquad (3.79)$$

Simple analytical expressions are also available for the concentration profiles

$$C_A(X,T) = 1 + (C_A(0) - 1)\,\mathrm{erfc}\left(\frac{X}{2\sqrt{T}}\right) \qquad (3.80)$$

$$C_B(X,T) = \frac{c_B^*}{c_A^*} + \left(C_B(0) - \frac{c_B^*}{c_A^*}\right)\mathrm{erfc}\left(\frac{X}{2\sqrt{d_B T}}\right) \qquad (3.81)$$

For multipulse voltammetric technique where a sequence of potential pulses $(E_1, E_2, \ldots E_n \ldots E_p)$ of the same duration (τ) is applied (as in cyclic staircase and square wave voltammetries, see Figure 1.10), the following expressions account for the dimensionless voltammetric response [12]

$$J(\theta) = \frac{1}{\sqrt{\pi T_{\text{pulse}}}} \sum_{n=1}^{P} \frac{C_A^{(n)}(0) - C_A^{(n-1)}(0)}{\sqrt{p - n + 1}} \tag{3.82}$$

where $T_{\text{pulse}} = D_A \tau / \varepsilon^2$ is the dimensionless duration of the pulses and $C_A^{(n)}(0)$ the dimensionless surface concentration of the reduced species in the n-th pulse

$$
\begin{aligned}
C_A^{(0)}(0) &= 1 \\
C_A^{(n)}(0) &= \frac{e^{\theta_n}\left(1 + \sqrt{d_B} c_B^*/c_A^*\right)}{\sqrt{d_B} + e^{\theta_n}} \quad (n > 0)
\end{aligned}
\tag{3.83}
$$

Note that the response in cyclic voltammetry can be calculated from (3.82) as the limit of the signal in staircase cyclic voltammetry (SCV) with a very small step potential (E_s <0.01 mV) [12]. It is also worth noting that the simulation of the SCV technique can be important itself in the analysis of experimental cyclic voltammograms obtained with modern digital instruments given that, instead of a linear ramp, these apply a sequence of pulses with a given E_s value. The minimum potential step can be restricted by equipment limitations such that significant differences between the SCV and CV responses can occur [13]. Under these conditions, a simple solution for the adequate analysis of the electrochemical response would be to adapt our simulation to the actual SCV perturbation applied by the potentiostat.[1]

Other testing criteria based on particular points of the voltammograms are possible. Thus, in cyclic voltammetry (see Figure 3.4), the dimensionless peak current [14–16] and potential of the forward scan (that is, the $J - \theta$ coordinates of the largest current recorded on the forward sweep) when the electrode reaction behaves reversibly take the values

$$|J_{\text{p}}| = 0.4463\sqrt{\sigma} \tag{3.84}$$

---

[1] Another popular solution for this problem is sampling the current in SCV not at the end of the pulse but at an appropriate time (sampling time, t_s) so that the SCV voltammogram is equivalent to the CV one. The optimal value for the sampling time depends on the experimental system (electrode kinetics, reaction mechanism, step potential, etc.) and it has been established for some frequent situations. For example, for a reversible E mechanism at a macroelectrode there is an acceptable agreement between SCV and CV for $E_s < 8$ mV and $t_s/t_p = 0.25$.

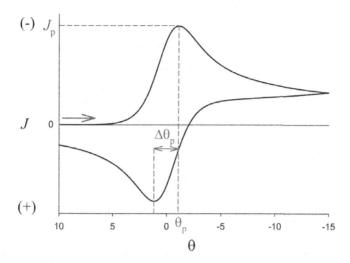

Fig. 3.4. Main characteristic points of the dimensionless cyclic voltammogram of a reversible one-electron reduction.

$$\theta_p = \theta_{1/2} \mp 1.109 \tag{3.85}$$

where the upper sign applies to an electro-reduction process, the lower one to an electro-oxidation and $\theta_{1/2}$ is the dimensionless reversible half-wave potential

$$\theta_{1/2} = 0.5 \ln(d_B) \tag{3.86}$$

Moreover, the peak-to-peak separation between the forward and backward peaks $(\Delta\theta_p)$ is close to

$$|\Delta\theta_p| \approx 2.3 \tag{3.87}$$

Thus, the accuracy of simulated cyclic voltammograms can be easily monitored from the error in the peak current and positions through the quantities

$$Q_{J_p} = \frac{J_{p,sim} - J_p}{J_p} \times 100 \tag{3.88}$$

$$Q_{\theta_p} = \theta_{p,sim} - \theta_p \tag{3.89}$$

$$Q_{\Delta\theta_p} = \frac{\Delta\theta_{p,sim} - \Delta\theta_p}{\Delta\theta_p} \times 100 \tag{3.90}$$

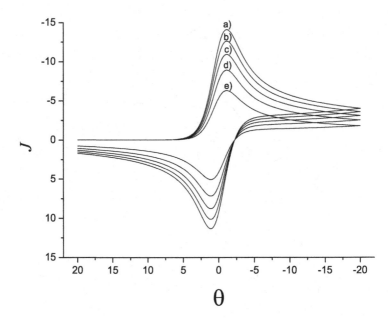

Fig. 3.5. Simulated cyclic voltammetry for a variety of scan rates and $\Delta X = 10^{-4}$, $\Delta\theta = 0.01$. $\sigma =$ (a) 10^3, (b) 8×10^2, (c) 6×10^2, (d) 4×10^2 and (e) 2×10^2.

By comparing the simulation's output with the values predicted by equations (3.84)-(3.87) for a range of values of σ, we can test to see if it is correct. Thus, for example, Figure 3.5 demonstrates how cyclic voltammetry varies with scan rate and Figure 3.6 demonstrates the agreement between the simulated results and the Randles–Ševčík equation.

In square wave voltammetry (see Figure 3.7), two simple and general criteria are the values of the dimensionless peak potential ($\theta_{\mathrm{p}}^{\mathrm{SW}}$) and half-peak width ($W_{1/2}$) for reversible electron transfers [12, 17, 18]

$$\theta_{\mathrm{p}}^{\mathrm{SW}} = \theta_{1/2} \tag{3.91}$$

$$W_{1/2} = \ln\left(\frac{1 + e^{2\theta_{\mathrm{SW}}} + 4e^{\theta_{\mathrm{SW}}} + \sqrt{\left(1 + e^{2\theta_{\mathrm{SW}}} + 4e^{\theta_{\mathrm{SW}}}\right)^2 - 4e^{2\theta_{\mathrm{SW}}}}}{1 + e^{2\theta_{\mathrm{SW}}} + 4e^{\theta_{\mathrm{SW}}} - \sqrt{\left(1 + e^{2\theta_{\mathrm{SW}}} + 4e^{\theta_{\mathrm{SW}}}\right)^2 - 4e^{2\theta_{\mathrm{SW}}}}}\right) \tag{3.92}$$

where $\theta_{1/2}$ is the dimensionless reversible half-wave potential and $\theta_{SW}\left(= \frac{F}{RT}|E_{SW}|\right)$ the dimensionless square wave amplitude. Hence, the

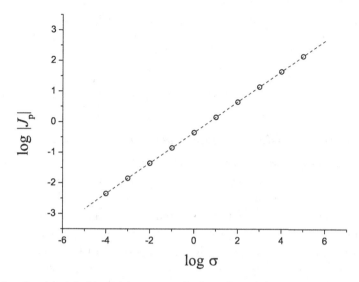

Fig. 3.6. Simulated J_p (circles) for a range of values of σ and that predicted by Eq. (3.84) (dashed line). $\Delta X = 10^{-4}$, $\Delta\theta = 0.01$.

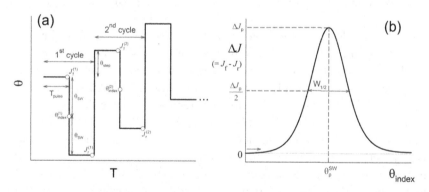

Fig. 3.7. (a) Parameters of the potential-time perturbation applied in square wave voltammetry; (b) main characteristic points of dimensionless square wave voltammograms.

simulated square wave voltammograms can be assessed through the values of

$$Q_{\theta_p^{SW}} = \theta_{p,sim}^{SW} - \theta_p^{SW} \qquad (3.93)$$

$$Q_{W_{1/2}} = \frac{W_{1/2,sim} - W_{1/2}}{W_{1/2}} \times 100 \qquad (3.94)$$

The SWV peak current ($\Delta J_{\text{plateau}}$) increases with the square wave amplitude and for very large values ($\theta_{\text{SW}} > 6$) the SWV signal shows a current *plateau* (instead of a peak), the height of which at macroelectrodes is given by [12]

$$\Delta J_{\text{plateau}} = 0.683 \left(1 + \frac{c_B^*}{c_A^*}\right) \frac{1}{\sqrt{T_{\text{pulse}}}} \qquad (3.95)$$

3.5.3. *Comprehensive simulation testing*

For the intensive validation of simulation programs, first it is necessary to identify all the parameters that affect the accuracy of the numerical solution. As mentioned above, in our context and for a given method, such parameters are primarily related to the refinement degree of the spatial and time grids. Thus, for equally spaced grids, two parameters can be identified: ΔX and ΔT (or $\Delta \theta$ in cyclic voltammetry). In Chapter 4, the use of unequally spaced grids will be considered, where additional parameters determine the grid refinement. Thus, in the case of an equally-spaced time grid combined with an expanding spatial grid (see Subsection 4.1.1), the effect of three simulation parameters are to be studied: ΔT (or $\Delta \theta$), the first spatial interval (h), and the grid expansion factor (ω_X).

Once all relevant *simulation parameters* have been identified, the accuracy of the simulated results as a function of them is monitored in order to reach the acceptable compromise between accuracy and runtime, the former being examined with indicators as those discussed in Subsections 3.5.1 and 3.5.2. To do this, we perform convergence studies under a variety of conditions of the *system and technique parameters* that determine the spatial and time scales where significant changes in concentration take place. For the electrochemical system simulated by the example program given in Subsection 3.4.1, this corresponds to the dimensionless scan rate σ: the faster the scan rate, the smaller the solution region where concentration gradients arise. Thus, the different fixed values of σ, ΔX and $\Delta \theta$ are varied and the accuracy of the simulated voltammograms and/or concentration profiles can be tested from their divergence from the known mathematical expressions or reference values.

The above procedure is illustrated in Figure 3.8 that shows the results of such a study in cyclic voltammetry for the cases where $\sigma = 1$ and $\sigma = 1000$ using a simulation program very similar to that given in Subsection 3.4.1. In this case, the error in the simulated peak height (Q_{J_p}) is considered and we have arbitrarily specified an accuracy tolerance of 0.1%, though ultimately it is up to the simulation user to judge what constitutes an acceptable level

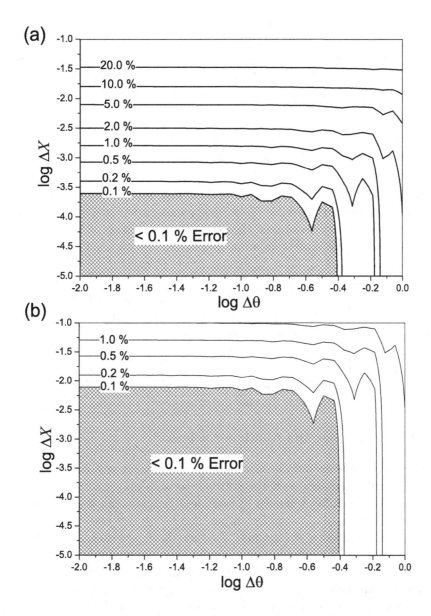

Fig. 3.8. Contour plots of percentage error of peak current from equation (3.84) (Q_{J_p}) for (a) $\sigma = 1000$; (b) $\sigma = 1$.

of accuracy. Any pair of values $(\Delta X, \Delta \theta)$ inside the shaded regions in the figure correspond to simulations with error $< 0.1\%$. As demonstrated, the lower the scan rate is, the greater ΔX may be while still achieving accurate results, though the dependence on $\Delta \theta$ does not seem to vary significantly. Consequently, any pair of values $(\Delta X, \Delta \theta)$ that give accurate results for one scan rate will also give equally or more accurate results for lower scan rates. From Figure 3.8(a), a reasonable choice of values would be $\Delta X = 2 \times 10^{-4}$ $[\log(\Delta X) \approx -3.7]$ and $\Delta \theta = 0.2$ $[\log(\Delta \theta) \approx -0.7]$ which would ensure error $< 0.1\%$ for $\sigma < 1000$ in a relatively short runtime.

Additional convergence studies are possible on the basis of different indicators $(Q_{\theta_p}, Q_{\Delta \theta_p}, \ldots)$ and each of them may lead to different optimum grids. For systematic and comprehensive simulation tests [7], it can be helpful to schedule and perform all the necessary simulations at different grids and system's conditions, computing and saving the values of the accuracy indicators of interest. Then, a common error analysis can be performed so that the simulation program is considered as suitable when all indicators fall below a given threshold according to our necessities. Note that such thresholds and indicators are to be selected as a function of the purpose of the simulations. For example, when analysing experimental voltammetric data, we are mainly interested in the current-potential response (rather than in the species concentration profiles), and 0.1% error in the simulated current and 0.1 mV in potential can be considered as suitable since they are well below the typical threshold of experimental accuracy.

3.6. Performance and Runtime Analysis

A program's runtime may be analysed in terms of the size of its inputs. For the example program above, the preamble section (prior to the main simulation loop) contains a number of variable definitions, simple calculations, the initialisation of a stream for output, and a loop that calculates the values of the modified γ_i' coefficients. The time taken to perform the variable definitions, calculations and stream initialisation is effectively constant for a given machine; it is dependent only on the specifics of the computer hardware and the compiler, and is independent of the program's inputs. The loop will take a time proportional to the size of the spatial grid, n (which is the number of loop iterations), plus some small constant time to initialise the loop variable. We may write that the runtime of the preamble section, τ_{pre}, is

$$\tau_{\mathrm{pre}} = B_1 + B_2 n \tag{3.96}$$

where B_1 is the constant time taken to initialise the variables (including the stream and the loop increment variable, i) and perform the calculations, and B_2 is the time taken per iteration of the γ_i' loop.

The main body of the simulation is a loop that iterates m times. It contains a number of constant-time operations and two separate loops which each iterate approximately n times. The runtime for the main body of the simulation is therefore

$$\tau_{\text{loop}} = m(B_3 + B_4 n) \tag{3.97}$$

and so the total runtime of the simulation is

$$\tau = B_4 mn + B_3 m + B_2 n + B_1 \tag{3.98}$$

It is difficult to determine *a priori* the value of the B constants and thus the absolute runtime for a given machine. Therefore, we will first examine how the runtime grows as a function of the input sizes, n and m. In our example above, as the sizes n and m increase, the first term in Eq. (3.98) dominates, and the other terms become less and less significant. Therefore in the limit of large n and m

$$\tau \approx B_4 mn \tag{3.99}$$

which in terms of the program's inputs is

$$\tau \approx B_4 \frac{12\sqrt{2}(\theta_v - \theta_i)^{3/2}}{\Delta X \Delta \theta \sqrt{\sigma}} \tag{3.100}$$

where the lower-order terms[2] are unimportant. B_4 can therefore be interpreted as the time taken by the computer to calculate the concentration at one space-time grid point. Its value depends on the exact structure of the program (how many operations are performed, what kind of operations, and in what order), the quality of the compiler (how efficiently the C++ instructions are translated into machine code) and the architecture of the machine on which the program is running (CPU clock speed, memory latency, etc.). For a particular program running on a particular machine, the

---

[2] In the analysis of algorithms [19], $\mathcal{O}$ notation (pronounced "big oh") is used to describe this relationship as the input sizes tend to large values. The notation, $\tau = \mathcal{O}(mn)$, means that the time as a function of m and n, $\tau(m,n)$, is bounded above by some constant multiple of mn for all sufficiently large values of m and n.

value may be easily determined; we explicitly record how long the program takes to run for given values of m and n and calculate B_4 from

$$B_4 = \frac{\tau}{mn} \tag{3.101}$$

where the runtime is recorded in seconds. In Windows, the time (in milliseconds) since the system started can be determined using the function `GetTickCount()`, and similar functions are available under other operating systems. Therefore to calculate the program's total runtime (in ms), we use this function once before the start of the program,

```
int startTime = GetTickCount();
```

and calculate the difference in time at the end, using the result to determine the value of B_4:

```
int timeElapsed = GetTickCount() - startTime;
double B4 = timeElapsed / (1000 * iterX * iterT);
```

Note that the `<windows.h>` header must be included in order to use this function. The simulation program does not run in isolation on a computer; the operating system must handle many tasks simultaneously and so the simulation may not be able to use the full resources of the computer for the whole time that it runs. Consequently, the runtime and thus the value of B_4 will vary slightly each time the program is run, even if the inputs are the same. Further, as discussed, the approximation in Eq. (3.101) will be less accurate for smaller values of m and n. Therefore in order to establish an accurate value for B_4, we vary the size of the inputs and then calculate the gradient $\frac{d\tau}{d(mn)}$ as demonstrated in Figure 3.9. On the particular machine used to run these simulations shown in this figure, $B_4 \approx 2.15 \times 10^{-8}$ s.

As discussed in Section 3.3, the general-purpose Gaussian elimination algorithm for inversing dense matrices takes on the order of n^3 operations, therefore a program that used that algorithm in place of the Thomas algorithm would take $\tau = \mathcal{O}(mn^3)$. Even for a small spatial grid with only 100 points, a program using the Gaussian algorithm would take 10,000 times longer than one using the Thomas algorithm! This highlights the importance of careful algorithm selection when performing numerical simulations.

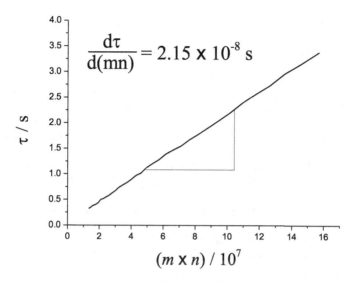

$$\frac{d\tau}{d(mn)} = 2.15 \times 10^{-8} \text{ s}$$

$(m \times n) / 10^7$

Fig. 3.9. Runtime, τ, against the size of the space-time grid, $m \times n$, for a series of simulations with varying scan rate, σ, and fixed values of $\Delta X = 10^{-4}$, $\Delta\theta = 0.01$ and $|\theta_v - \theta_i| = 40$.

3.6.1. *Memory requirements*

The variables used in the program must be stored in the computer's main memory (RAM). While a number of variables are assigned during the course of the program, the overwhelming majority of the memory used is consumed by the vectors that are used to store C_i^k, δ_i^k, $\delta_i'^k$ and γ_i', each of which are of size n. From Eq. (3.98), if $\theta_v = -20$ and $\theta_i = 20$

$$n = \frac{53.67}{\Delta X \sqrt{\sigma}} \tag{3.102}$$

Each value in each of the arrays is of type **double**; these occupy 8 bytes (64 bits) of memory each. If we assume a value of $\Delta X = 10^{-4}$, then the memory required in megabytes (where 1 MB = 1024^2 bytes) is

$$\text{Size} \approx 4 \times 8 \times \frac{537000}{\sqrt{\sigma}} \times \frac{1}{1024^2} \approx \frac{16.4}{\sqrt{\sigma}} \text{ MB} \tag{3.103}$$

If we assume a maximum available memory size of 4 GB, then the smallest value of σ that we could possibly use with our current simulation would be $\approx 10^{-5}$, though the runtime would be prohibitive.

References

[1] K. W. Morton and D. F. Mayers, *Numerical Solution of Partial Differential Equations: An Introduction.* (Cambridge University Press, Cambridge, 2005), 2nd edition.

[2] G. D. Smith, *Numerical Solution of Partial Differential Equations: Finite Difference Methods.* (Oxford University Press, Oxford, 1985), 3rd edition.

[3] W. H. Press, S. A. Teukolsky, W. T. Vetterling, and B. P. Flannery, *Numerical Recipes.* (Cambridge University Press, Cambridge, 2007), 3rd edition.

[4] R. G. Compton, M. E. Laing, D. Mason, R. J. Northing, and P. R. Unwin, Rotating disk electrodes: The theory of chronoamperometry and its use in mechanistic investigations., *Proc. R. Soc. Lond. A Math. Phys. Sci.* **418**, 113–154, (1988).

[5] D. Britz and J. Strutwolf, *Digital Simulation in Electrochemistry. 4th ed.* (Springer, Berlin, 2016).

[6] D. J. Gavaghan, How accurate is your two-dimensional numerical simulation? Part 1. An introduction, *J. Electroanal. Chem.* **420**, 147–158, (1997).

[7] E. Kätelhön and R. G. Compton, Testing and validating electroanalytical simulations, *Analyst.* **140**(8), 2592–2598, (2015).

[8] K. B. Oldham and S. W. Feldberg, Principle of unchanging total concentration and its implications for modeling unsupported transient voltammetry, *The Journal of Physical Chemistry B.* **103**(10), 1699–1704, (1999).

[9] P. A. Kottke and A. G. Fedorov, Generalized principles of unchanging total concentration, *The Journal of Physical Chemistry B.* **109**(35), 16811–16818, (2005).

[10] E. J. Dickinson, J. G. Limon-Petersen, and R. G. Compton, The electroneutrality approximation in electrochemistry, *Journal of Solid State Electrochemistry.* **15**(7-8), 1335–1345, (2011).

[11] A. Molina and E. Laborda, Detailed theoretical treatment of homogeneous chemical reactions coupled to interfacial charge transfers, *Electrochimica Acta.* **286**, 374–396, (2018).

[12] A. Molina and J. González, *Pulse Voltammetry in Physical Electrochemistry and Electroanalysis.* (Springer, Berlin, 2016).

[13] M. Seralathan, R. Osteryoung, and J. Osteryoung, Comparison of linear sweep and staircase voltammetries using Walsh series, *J. Electroanal. Chem.* **214**, 141–156, (1986).

[14] R. G. Compton and C. E. Banks, *Understanding Voltammetry, 3rd ed.* (World Scientific Press, Singapore, UK, 2018).

[15] J. E. B. Randles, Cathode-ray polarograph. II. Current-voltage curves., *Trans. Faraday Soc.* **44**, 327–338, (1948).

[16] A. Ševčík, Oscillographic polarography with periodical triangular voltage., *Collect. Czech. Chem. Commun.* **13**, 349–377, (1948).

[17] J. J. O'Dea, J. Osteryoung, and R. A. Osteryoung, Theory of square wave voltammetry for kinetic systems, *Analytical chemistry.* **53**(4), 695–701, (1981).

[18] V. Mirceski, S. Komorsky-Lovric, and M. Lovric, *Square-wave voltammetry: theory and application.* (Springer Science & Business Media, 2007).

[19] R. R. T. Cormen, C. Leiserson and C. Stein, *Introduction to Algorithms.* (MIT Press, 2009), 3rd edition.

Chapter 4

Diffusion-Only Electrochemical Problems in One-Dimensional Systems

In the previous chapter finite difference methods were introduced for one of the simplest situations from a theoretical point of view; cyclic voltammetry of a reversible E mechanism (i.e., charge transfer without chemical complications) at a planar electrode and with equal diffusion coefficients for the electroactive species. However, electrochemical systems are typically more complex and some refinements must be introduced in the numerical methods for adequate modelling.

In this chapter, some simple extensions will be considered that will enable the simulation of E mechanisms when the electrode reaction is kinetically slow (non-reversible processes) and when the diffusivities of the electroactive species are different. Moreover, advanced discretisations for the spatial and temporal grids will be introduced in order to increase the efficiency of the simulation.

Finally, the cases of spherical, hemispherical and cylindrical electrodes will be tackled, which cover the use of wire electrodes, mercury drops and microhemispheres, and liquid-liquid interfaces. These geometries enable us to introduce the effects due to convergent diffusion on the mass transport and voltammetric response. Moreover, as in the case of planar electrodes, because of the symmetry of the mass transfer field the problems can each be reduced to only one dimension: the distance to the electrode surface in the normal direction.

4.1. Unequally Spaced Grids

Figure 4.1 shows the variation of the reactant concentration with the distance to the electrode, x, (that is, the concentration profile) in a Cottrell experiment under semi-infinite linear diffusion. A large overpotential is applied to the working electrode such that the reactant species is consumed

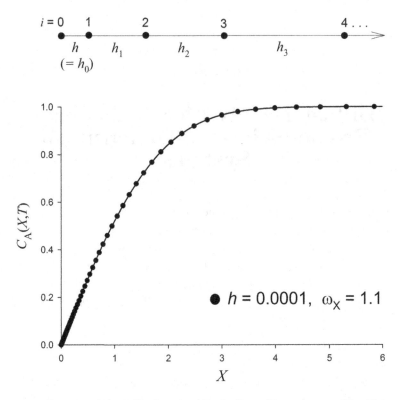

Fig. 4.1.   Concentration profile of species A in the Cottrell experiment at $T = T_{\max} = 1$. The solid line corresponds to the profile obtained with a uniform grid with $h = 0.0001$; the circle points indicate the position of the spatial nodes when an expanding grid is used with $h = 0.0001$ and $\omega_X = 1.1$. $\Delta T = 0.01$.

very rapidly on the electrode surface $(c_A (x = 0) = 0)$. This creates a concentration gradient in the vicinity of the electrode and consequently the diffusion of species towards the electrode surface such that the depletion layer extends towards the bulk solution as time proceeds. As a result, the concentration profile shown in Figure 4.1 develops. This is representative of the concentration profiles of the reactant in typical voltammetric experiments where the steepest concentration gradient occurs next to the electrode surface. Therefore, the interval between grid points in this region must be small enough to minimise the error of the finite difference approximation. On the other hand, the concentration profiles of the electroactive species are only slightly affected in the regions of the solution farther away from the electrode surface.

Considering the above aspects, it is beneficial to establish a grid that mimics the concentration profile[1] and provides a high number of points near the electrode surface but minimises the total number, which makes the simulation process more efficient. With this aim, unequally spaced grids or transformation of the spatial coordinates can be employed, the former being preferred in the simulation of electrochemical experiments [3].

In the same way, the time grid can be designed to increase the computational efficiency of the numerical simulation. Following the idea discussed for the spatial grid, the number of timesteps must be higher at those times when the changes in concentration with time are more significant, whereas the time interval can be longer under conditions where the time dependence of concentration is smaller. This strategy is interesting in the case of chronoamperometry and pulse techniques where abrupt changes in concentration take place typically at the beginning of the application of the potential pulses.

4.1.1. *Expanding spatial grid*

The distribution of nodes of the spatial grid is adjusted to concentrate a higher number of points in the regions of the concentration profile where changes are more significant, as can be seen in the upper scheme shown in Figure 4.1. This region can be defined by the mass transport of the electroactive species or by the so-called *reaction layer* (see Chapter 5) when the electron transfer is complicated by coupled chemical reactions in solution.

Exponentially expanding grids are widely employed in the simulation of electrochemical experiments such that the distance between consecutive points of the grid expands exponentially according to the following definition:

$$h_i = X_{i+1} - X_i = h\omega_X^i \tag{4.1}$$

where $0 \leq i < (n - 1)$, h_i is the distance between the points i and $i + 1$, h the distance between the first two points of the spatial grid and ω_X the expansion factor (see Figure 4.1). In order to ensure that the last point of

---

[1] In the case of multi-E mechanisms, kinetic fronts associated with comproportionation reactions can take place in solution instead of next to the electrode surface. In this particular case, very dense grids or automatically adaptive grids are employed for accurate description of the concentration profiles (see Chapter 6) [1, 2].

the grid is situated at a distance $\geq 6\sqrt{T_{\max}}$ from the electrode surface, the total number of points is adjusted.

In terms of the programming, it is necessary to calculate the position of every grid point X_i at the start of a simulation. The following code fragment can be used for such a calculation:

```
//Expanding grid
double omega = 1.1;
double h = 1e-4;

std::vector<double> X;

X.push_back(0.0);
while(X.back() < maxX)  // where maxX = 6*sqrt(maxT)
{
    X.push_back(X.back() + h);
    h *= omega;
}

int n = X.size();
```

The position of every point in the spatial grid is stored in a vector, and the number of spacesteps, n, is now equal to the number of elements in this vector (rather than being calculated ahead of time).

After introducing unequal spatial intervals, the finite differences must be reformulated. Following the central three-point approximation introduced in Chapter 3, the second derivative can be approximated as

$$\frac{\partial^2 C}{\partial X^2} \approx \frac{\frac{C_{i+1}-C_i}{\Delta X_+} - \frac{C_i-C_{i-1}}{\Delta X_-}}{\frac{1}{2}[\Delta X_+ + \Delta X_-]} \tag{4.2}$$

where $\Delta X_+ = X_{i+1} - X_i (= h_i)$, $\Delta X_- = X_i - X_{i-1} (= h_{i-1})$ so that Fick's second law for linear diffusion, employing the fully implicit method (backward implicit, BI) introduced in Chapter 3, is given by

$$\frac{C_i^k - C_i^{k-1}}{\Delta T} = \frac{\frac{C_{i+1}^k-C_i^k}{\Delta X_+} - \frac{C_i^k-C_{i-1}^k}{\Delta X_-}}{\frac{1}{2}[\Delta X_+ + \Delta X_-]} \tag{4.3}$$

which can be rearranged as

$$
C_i^{k-1} = - \left[\frac{2\Delta T}{\Delta X_-^2 + \Delta X_+ \Delta X_-} \right] C_{i-1}^k
$$
$$
+ \left[\frac{2\Delta T}{\Delta X_+^2 + \Delta X_+ \Delta X_-} + \frac{2\Delta T}{\Delta X_-^2 + \Delta X_+ \Delta X_-} + 1 \right] C_i^k \qquad (4.4)
$$
$$
- \left[\frac{2\Delta T}{\Delta X_+^2 + \Delta X_+ \Delta X_-} \right] C_{i+1}^k
$$

From the above equation, the coefficients α_i, β_i and γ_i for the Thomas algorithm (see Chapter 3) can be identified

$$
\alpha_i = - \frac{2\Delta T}{\Delta X_-^2 + \Delta X_+ \Delta X_-} \qquad (4.5)
$$

$$
\beta_i = \frac{2\Delta T}{\Delta X_+^2 + \Delta X_+ \Delta X_-} + \frac{2\Delta T}{\Delta X_-^2 + \Delta X_+ \Delta X_-} + 1 \qquad (4.6)
$$

$$
\gamma_i = - \frac{2\Delta T}{\Delta X_+^2 + \Delta X_+ \Delta X_-} \qquad (4.7)
$$

Note that the above coefficients reproduce those given in Chapter 3 for equally spaced grids by making $\omega_X \to 1$ so that $\Delta X_+ = \Delta X_- = \Delta X$.

When $\omega_X \neq 1$, the ΔX_- and ΔX_+ values are different for each point of the grid such that coefficients α_i, β_i and γ_i are no longer common and their values depend on the absolute position in the grid (i.e., on the i value). Consequently, the Thomas algorithm coefficients must be calculated for each spatial point before starting the simulation:

```
// Calculate Thomas coefficients
std::vector<double> alpha(n-1, 0.0);
std::vector<double> beta(n-1, 0.0);
std::vector<double> gamma(n-1, 0.0);

for(int i=1; i<n-1; i++)
{
    double delX_m = X[i] - X[i-1];
    double delX_p = X[i+1] - X[i];

    alpha[i] = -(2.0* deltaT) / ( delX_m * (delX_m + delX_p) );
    gamma[i] = -(2.0* deltaT) / ( delX_p * (delX_m + delX_p) );
    beta[i] = 1.0 - alpha[i] - gamma[i];
}
```

The other change required (compared to the program listed in Subsection 3.4.1) is to subscript all further occurrences of `alpha`, `beta` and `gamma` with `[i]`. Note also that a two-point approximation should be sufficient for the current calculations due to the small size of the first spacestep when using an expanding grid:

$$J^k = -\left(\frac{C_1^k - C_0^k}{h}\right) \tag{4.8}$$

The distribution of points in the spatial grid can be adjusted by means of the parameters $h(= h_0)$ and ω_X. For example, the distribution of nodes corresponding to $\omega_X = 1.1$ and $h = 0.0001$ in the Cottrell experiment are shown in Figure 4.1. The expanding grid enables us to have a very dense grid $(X_1 - X_0 = 0.0001)$ next to the electrode surface and, at the same time, to cover all the simulation space from $X = 0$ to $6\sqrt{T_{\max}}$ with only 93 points (instead of 10^4!). Hence, this implementation is simple and, in return, the number of points is greatly reduced such that the use of exponentially expanding grids significantly improves the simulation efficiency.

The optimal h and ω_X values must be determined by convergence study to ensure a compromise between accuracy and computation runtime. Thus, small h and ω_X values diminish the error of the finite difference approximation of the concentration derivatives but also involve a large number of points and thus high consumption of computing resources. On the other hand, the use of large h and ω_X values reduces the number of points in the grid (and so the CPU time) but can give rise to unacceptable errors.[2]

Other alternatives arise from the use of patching schemes that combine uniform, expanding and compressing grids. For example, assuming that the main changes in concentration take place next to the electrode surface, a dense, uniform grid in that region $(X \leq X_s)$ can be employed followed by an expanding grid towards the bulk solution:

$$\begin{aligned} \text{for} \quad & 0 \leq X \leq X_s: \quad & h_i &= h \\ \text{for} \quad & X_s < X: \quad & h_i &= h\omega_X^{i-(Xs/h)+1} \end{aligned} \tag{4.9}$$

where X_s can be defined, for example, as a multiple of the diffusion or reaction layer thickness.

Patching schemes are helpful in situations where there are singularities in the concentration profiles in certain regions, such as the edge of disc

---

[2] Larger expansion factors can be used with accurate results by employing more points in the approximation of the derivatives (higher-order approximations) [4].

microelectrodes (see Chapter 9) or the boundary of the diffusion domain in thin-film voltammetry, amalgamation processes and electrode arrays (see Chapter 10). In the latter case, the domain is confined to a distance d from the electrode surface at which a zero-flux condition applies and a well-defined grid may be necessary around this point such that the discretisation of the corresponding spatial derivative is accurate. A simple solution is to stop the expansion when the points are approaching the end of the diffusion domain and turn to a dense, uniform grid. Another alternative arises from the use of expanding-compressing schemes such that

$$
\begin{aligned}
\text{for} \quad 0 \leq X \leq X_s: \quad & h_i = h\omega_X^i \\
\text{for} \quad X_s < X: \quad & h_i = h\omega_X^{(n-2-i)}
\end{aligned}
\tag{4.10}
$$

with $0 \leq i < (n-1)$ where n is the total number of points of the spatial grid. This will be discussed in more detail when tackling the simulation of disc electrodes in Chapter 9.

4.1.2. *Expanding time grid: Chronoamperometry and pulse techniques*

Applying the philosophy of expanding grids to time discretisation, one can reason that the interval between timesteps can also be adjusted by placing more points at time values where changes in concentration are more rapid. This approach is particularly useful for the simulation of chronoamperometry and potential pulse techniques.

When applying a constant potential pulse, rapid changes in concentration take place at the beginning of the pulse while changes are much less significant at long times. This behaviour can be followed through the time variation of the surface concentration gradient, which is proportional to the current passing through the cell that changes rapidly with time after the potential jump as shown in the chronoamperogram of Figure 4.2. Therefore, small time intervals (ΔT) are necessary at short times in order to ensure good accuracy of the finite difference approximation. On the other hand, larger ΔT values can be used at long times where the variation with time of the concentration profiles and current response is slower.

Following the patching scheme described above for the spatial discretisation, the beginning of each potential pulse can be simulated through a uniform, dense time grid with the interval ΔT:

$$
\Delta T^k = \Delta T, \quad 0 < T \leq T_s \quad (k = 1, 2, \ldots, T_s/\Delta T)
\tag{4.11}
$$

Fig. 4.2. Variation of the current response with time in the Cottrell experiment with $T_{\max} = 1$. The line corresponds to the exact solution (the Cottrell equation (3.72)) and the circle points indicate the position of the timesteps when an expanding grid is used with $\Delta T = 0.01$, $\omega_T = 1.02$ (and $T_s = 0$). $h = 0.0001$ and $\omega_X = 1.1$.

then followed by an exponentially expanding time grid [5]:

$$\Delta T^k = \Delta T \omega_T^{k-(T_s/\Delta T)} \, , \qquad T > T_s \quad (k > T_s/\Delta T) \qquad (4.12)$$

where $\Delta T^k = T^k - T^{k-1}$. Note that in pulse techniques we are mainly interested in the value of the current at the end of each pulse; therefore, when unevenly spaced time grids are employed, the duration of the last interval can be adjusted (rather than following the expansion given by (4.12)) so that the last point corresponds to $T = T_{\max}$. The ω_T value enables us to minimise the number of time points (m) in the region $T_s < T \le T_{\max}$. Again, the ΔT and ω_T values are selected such that a suitable compromise between accuracy and runtime is made.

It is worth highlighting that, analogously to the case of non-uniform space grids, the use of unequal time intervals introduces a new dependence in the coefficients for the Thomas algorithm (α, β and γ given by Eqs. (4.5)–(4.7) with $\Delta T = \Delta T^k$) which are now different for each timestep for $T > T_s$ since the time interval ΔT^k is different for each k value. Consequently, the Thomas coefficients must be calculated for every timestep and so there are situations where the use of non-uniform time grids may not be advantageous in terms of efficiency.[3]

4.1.3. *Example program*

The following C++ program simulates the chronoamperometry of a one-electron reversible reduction at a macroelectrode with equal diffusion coefficients, employing unequally-spaced spatial and time grids. First, the values of the dimensionless applied potential (θ) and pulse duration ($T_{\max} = T_{\text{pulse}} = D\tau/\varepsilon^2$ where τ is the dimensioned duration of the pulse) are specified, as well as the parameters that define the exponentially expanding spatial grid (ω_X and h, Eq. (4.1)) and the time grid (ω_T, ΔT and T_s, Eqs. (4.11) and (4.12)).[4] Then, the Thomas coefficients are calculated for *each* spatial point and, for $T > T_s$, in *each* timestep.

Note that this program can be taken as a basis to develop simulations of multipulse techniques with an additional loop where each iteration corresponds to a potential pulse and the θ value is varied according to the corresponding potential-time perturbation. In such case, it is worth recalling that $T_{\max}$ should be referred to the *total* duration of the experiment, otherwise the simulation space ($0 < X < 6\sqrt{T_{\max}}$) may not be large enough to account for semi-infinite diffusion conditions. Thus, for example, if n pulses of the same duration τ are applied, then $T_{\max} = n \times D\tau/\varepsilon^2 = n \times T_{\text{pulse}}$.

· With respect to the program in Subsection 3.4.1, a couple of new functionalities have been added, which are useful for the assessment of the performance of the simulation: The determination of the simulation runtime (through the function `clock()`) and of the error of the simulated current (Q_J, Eq. (3.75)), as well as the output of the concentration profile of the reactant at the end of the experiment.

```
#include <fstream> // Header for file output
#include <vector> // Header for vectors
#include <cmath> // Header for sqrt(), fabs()
#include <ctime> // Header for calculating the runtime
#include <iostream> // Header for output on screen
```

---

[3] Time integration methods more advanced than the one considered in this book (backward implicit, BI) can also be employed. Indeed, the Crank–Nicholson and high-order extrapolation methods [6] have proven to enable the reduction of the number of timesteps (and even improve the accuracy of the simulation) with respect to BI [4].

[4] Equally spaced grids can be employed by setting $\omega_X = 1$ and/or $\omega_T = 1$ and $T_s = T_{\max}$.

```cpp
int main()
{
    // Specify simulation parameters
    double Theta = -15.0;// Dimensionless applied potential
    double Tmax = 1.0;// Dimensionless duration of the potential pulse

    double omegaX = 1.05;
    double h = 1.e-4;

    double omegaT = 1.015;
    double deltaT = Tmax / 4000.0;
    double Ts = Tmax / 2.0;

    clock_t runtime;
    runtime = clock();

    //Calculate other parameters
    // Expanding spatial grid
    double maxX = 6.0*sqrt(Tmax);
    std::vector<double> X;
    X.push_back(0.0);
    while(X.back() < maxX)
    {
        X.push_back(X.back() + h);
        h *= omegaX;
    }
    int n = X.size();// number of spacesteps

    //Expanding time grid
    std::vector<double> T;
    T.push_back(0.0);
    while(T.back() < Tmax) // where maxX = 6*sqrt(Tmax)
    {
        T.push_back(T.back() + deltaT);
        if(T.back() > Ts) deltaT *= omegaT;
    }
    int m = T.size();// number of timesteps
    T[m-1] = Tmax;

    // Create containers
    std::vector<double> alpha(n-1, 0.0);
    std::vector<double> beta(n-1, 0.0);
    std::vector<double> gamma(n-1, 0.0);
    std::vector<double> g_mod(n-2, 0.0);
    std::vector<double> delta(n-1, 1.0);
    std::vector<double> d_mod(n-1, 0.0);
    std::vector<double> C(n, 1.0); // concentration profile

    // Calculate Thomas coefficients for T < Ts
    double delT = T[1]-T[0];
```

```cpp
std::vector<double> delX(n, 0.0);
delX[0] = X[1] - X[0];
for(int i=1; i<n-1; i++)
{
    delX[i] = X[i+1] - X[i];

    alpha[i] = -(2.0* delT) / ( delX[i-1] * (delX[i-1] + delX[i]) );
    gamma[i] = -(2.0* delT) / ( delX[i] * (delX[i-1] + delX[i]) );
    beta[i] = 1.0 - alpha[i] - gamma[i];
}

// Modified gamma coefficients for T < Ts
g_mod[0] = 0.0; // boundary condition
for(int i=1; i<n-2; i++) g_mod[i] = gamma[i]
                        / (beta[i] - g_mod[i-1] * alpha[i]);


// Open file to output chronoamperogram and concentration profile
std::ofstream Chrono("Chrono_Output.txt");
std::ofstream Profile("Profile_Output.txt");


// BEGIN SIMULATION
for(int k=1; k<m; k++)
{
    // Calculate Thomas coefficients for T > Ts
    if(T[k] > Ts){
      for(int i=1; i<n-1; i++)
      {
        delT = T[k] - T[k-1];
        alpha[i] = -(2.0* delT) / ( delX[i-1] * (delX[i-1] + delX[i]) );
        gamma[i] = -(2.0* delT) / ( delX[i] * (delX[i-1] + delX[i]) );
        beta[i] = 1.0 - alpha[i] - gamma[i];
      }
    // Modified gamma coefficients for T > Ts
    g_mod[0] = 0.0; // boundary condition
    for(int i=1; i<n-2; i++) g_mod[i] = gamma[i]
                        / (beta[i] - g_mod[i-1] * alpha[i]);

    }

    // Forward sweep - create modified deltas
    d_mod[0] = 1.0 / (1.0 + exp(-Theta));
    for(int i=1; i<n-1; i++) {
        d_mod[i] = ( delta[i] - d_mod[i-1]*alpha[i] )
            / ( beta[i] - g_mod[i-1] * alpha[i] );
    }

    // Back Substitution
    C[n-1] = 1.0;
```

```
for(int i=n-2; i>=0; i--) {
    C[i] = d_mod[i] - g_mod[i] * C[i+1];

    delta[i] = C[i];
}

//Output current (and error)
double flux = -(C[1] - C[0])/(X[1] - X[0]);

double reference = - 1.0 / sqrt(M_PI*T[k]);
double QJ = (flux - reference) / reference *100.0;
Chrono << T[k] << "\t" << flux << "\t" << QJ << "\n";
}
// END SIMULATION

runtime = (float)(clock() - runtime)/CLOCKS_PER_SEC*1e3;
std::cout << "Runtime = " << runtime << " ms" << std::endl;

//Output the concentration profile at the end of the pulse
for(int i=0; i<n; i++) Profile << X[i] << "\t" << C[i] << "\n";

}
```

4.2. Finite Electrode Kinetics

So far we have considered that the electrode reaction is very fast (reversible) such that equilibrium conditions apply for the surface concentrations of the electroactive species A and B: $C_{A,0} = e^\theta C_{B,0}$. Nevertheless, electron exchange reactions can involve significant changes of molecular configuration such that the reaction is slow and electrochemical equilibrium is not achieved in the time scale of the experiment. Therefore it is important to investigate the general case of finite-kinetic electron transfer processes, considering the reversible case as a particular situation where the electrode kinetics are very fast with respect to the mass transport and the time scale of the experiment. Thus, the net reaction rate is given by

$$\frac{I}{FA} = -D_A \left(\frac{\partial c_A(x,t)}{\partial x} \right)_{x=0} = -\left[k_{red} c_A(0,t) - k_{ox} c_B(0,t) \right] \qquad (4.13)$$

where k_{red} and k_{ox} are the reduction and oxidation rate constants, the value of which depend on the applied potential and can be related through the Nernst equation:

$$k_{ox} = k_{red} e^\theta \qquad (4.14)$$

4.2.1. *The Butler–Volmer model*

There are different formalisms for the modelling of the potential dependence of the rate constants. The empirical Butler–Volmer (BV) model has been the most widely used over a number of years in electrochemistry due to its simplicity and successful quantitative description of a vast number of electrochemical systems (in the absence of bonds being broken or formed). According to the BV model, the rate constants show a simple exponential dependence with the applied potential according to the following expressions:

$$k_{red}^{BV} = k_0 e^{-\alpha\theta}$$
$$k_{ox}^{BV} = k_0 e^{(1-\alpha)\theta}$$

$$(4.15)$$

so that the electrode kinetics are parameterised as a function of the formal potential E_f^0 (included in the dimensionless parameter θ), the standard heterogeneous rate constant k_0 (m s^{-1}) and the charge transfer coefficient α. The k_0 value corresponds to the value of the rate constants at the formal potential ($k_{red}(E = E_f^0) = k_{ox}(E = E_f^0) = k_0$) and it is considered for the classification of electrode processes as reversible, quasi-reversible or irreversible. The dimensionless parameter α can take values in the range $0 \leq \alpha \leq 1$ (although typical values are in the range $0.3 < \alpha < 0.7$) and can be interpreted qualitatively as an indicator of the position of the transition state in the reaction coordinate. Thus, the configuration of the transition state in a reduction process is reactant-like when $\alpha > 0.5$ and product-like when $\alpha < 0.5$.

Next, we deal with the implementation of the new surface boundary condition in our simulation. As mentioned before, when the diffusion coefficients of both electroactive species are the same ($D_A = D_B = D$) the total concentration of the species at any point of the solution and at any time is constant: $c_A(x,t) + c_B(x,t) = c_A^* + c_B^*$. Along with Eq. (4.14), this enables us to rewrite Eq. (4.13) as a function of only the concentration of species A in the followng way:

$$D\left(\frac{\partial c_A(x,t)}{\partial x}\right)_{x=0} = k_{red}\left\{c_A(0,t) - [c_A^* + c_B^* - c_A(0,t)]e^{\theta}\right\} \quad (4.16)$$

Attending to the dimensionless variables, $X = x/\epsilon$ and $T = Dt/\epsilon^2$, defined in Chapter 2, we can express it as

$$\left(\frac{\partial C_A(X,T)}{\partial X}\right)_{X=0} = f(\theta)\frac{k_0\epsilon}{D}\left\{C_A(0,T) - [1 + C_B^* - C_A(0,T)]e^{\theta}\right\}$$

$$(4.17)$$

where $C_B^* = c_B^*/c_A^*$ and the form of $f(\theta)$ depends on the kinetic model employed:

$$f_{BV}(\theta) = e^{-\alpha\theta} \tag{4.18}$$

$$f_{MH}(\theta) = \frac{I(\theta, \Lambda, \gamma)}{I(0, \Lambda, \gamma)} \tag{4.19}$$

where f_{BV} corresponds to the Butler–Volmer model and f_{MH} to the Marcus–Hush model (see below).

Note that a new dimensionless variable appears that accounts for the kinetics of the electrode reaction:

$$K_0 = \frac{k_0 \epsilon}{D} \tag{4.20}$$

such that the smaller the K_0 value, the slower the electrode process (i.e., the more irreversible the system).

After discretisation of the problem using a two-point approximation for the surface gradient we obtain

$$\frac{C_{A,1}^k - C_{A,0}^k}{h} = f(\theta^{(k)})K_0 \left[C_{A,0}^k - \left(1 + C_B^* - C_{A,0}^k\right) e^{\theta^{(k)}} \right] \tag{4.21}$$

that can be rearranged as

$$C_{A,0}^k \left[1 + hf(\theta^{(k)})K_0 \left(1 + e^{\theta^{(k)}}\right)\right] - C_{A,1}^k = hf(\theta^{(k)})K_0(1+C_B^*)e^{\theta^{(k)}} \tag{4.22}$$

where h is the first spatial interval: $h = X_1 - X_0$. Therefore, the coefficients for the Thomas algorithm corresponding to the first point of the spatial grid $(i = 0)$ are now given by

$$\alpha_0 = 0 \tag{4.23}$$

$$\beta_0^{(k)} = 1 + hf(\theta^{(k)})K_0 \left(1 + e^{\theta^{(k)}}\right) \tag{4.24}$$

$$\gamma_0 = -1 \tag{4.25}$$

$$\delta_0^{(k)} = hf(\theta^{(k)})K_0(1 + C_B^*)e^{\theta^{(k)}} \tag{4.26}$$

For simulation purposes, altering a program that uses Nernstian equilibrium to use instead Butler–Volmer kinetics is a simple matter of changing the values of these coefficients (noting that $f(\theta^{(k)}) = e^{-\alpha\theta^{(k)}}$ for BV). Once the concentration profile of species A is calculated, that of species B can be obtained immediately from $C_{B,i}^{(k)} = 1 + C_B^* - C_{A,i}^{(k)}$.

4.2.2. Testing Simulations: Specific tests

A simple analytical solution is available for the chronoamperometric response of non-reversible electrode processes of any kinetics at macro-electrodes

$$J(T) = -\sqrt{\frac{d_B}{T}} \left(\frac{1 - e^\theta c_B^*/c_A^*}{\sqrt{d_B} + e^\theta} \right) \frac{\chi}{2} \exp\left(\chi/2\right)^2 \text{erfc}\left(\chi/2\right) \quad (4.27)$$

where

$$\chi = 2K_0 f\left(\theta\right) \sqrt{\frac{T}{d_B}} \left(\sqrt{d_B} + e^\theta \right) \quad (4.28)$$

With regard to multipulse techniques, the effect of the dimensionless kinetic parameter K_0 on the cyclic voltammetry and square wave voltammetry of an electro-reduction process are shown in Figure 4.3 based on the BV model. In both techniques, as the K_0 value decreases and the process becomes less reversible, the peak current decreases and the peak potentials move away from the formal potential, giving rise to the increase of the CV peak-to-peak separation (which is larger than the value of $2.218 \, RT/\text{F}$ for reversible processes).

For large K_0 values, the simulated results can be tested according to the expressions given in Subsubsection 3.5.2.2 for reversible processes. For slow kinetics (that is, fully irreversible processes), the BV model predicts that the peak height in CV and SWV are independent of K_0, while their position continues to move away from E_f^0 as K_0 decreases. In the case of CV, $J_{p,irre}$ scales with the square root of the scan rate according to the following expression for a reduction

$$|J_{p,irre}| = 0.4958\sqrt{\alpha \, \sigma} \quad (4.29)$$

and

$$|J_{p,irre}| = 0.4958\sqrt{(1 - \alpha)\,\sigma} \quad (4.30)$$

for an oxidation. The dimensionless peak potentials are given by

$$\theta_{\text{p,irre}} = -\frac{1}{\alpha} \left[0.78 - \ln\left(\frac{K_0}{\sqrt{\alpha\sigma}} \right) \right] \quad (4.31)$$

and

$$\theta_{\text{p,irre}} = \frac{1}{(1 - \alpha)} \left[0.78 - \ln\left(\frac{K_0}{\sqrt{(1 - \alpha)\sigma}} \right) \right] \quad (4.32)$$

for a reduction and an oxidation, respectively.

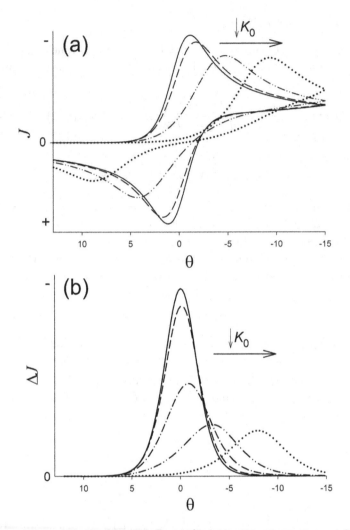

Fig. 4.3.   Effect of the electrode kinetics on the (a) cyclic voltammetry and (b) square wave voltammetry of an electro-reduction process at a macroelectrode. $\alpha = 0.5$, $C_B^* = 0$.

4.2.3. *The Marcus–Hush model*

The BV model has been very successful in the parameterisation and classi-fication of the kinetics of systems where the electroactive species diffuse in the electrolytic solution instead of being attached to the electrode surface. On the other hand, for the case of electroactive monolayers, significant de-viations have been reported with respect to BV predictions [7]. Moreover, the BV model is very limited with respect to understanding the connection between the physico-chemical properties of the system and its kinetics and to making predictions.

To fill the above gaps, microscopic kinetic models must be employed, the Marcus–Hush model (MH) being the preferred choice. The MH model introduces the concept of *reorganisation energy*, λ (eV), which corresponds to the energy required to distort the atomic configurations of the reactant molecule and its solvation shell to those of the product in its equilibrium configuration. Therefore, the λ value has a clear connection with the properties of the system and it enables us *a priori* estimation of the reaction rate.

According to the asymmetric version of the MH model (aMH)[5] the expressions for the oxidation and reduction rate constants are given by

$$k_{\text{red}}^{\text{MH}} = k_0 \frac{I(\theta,\Lambda,\gamma)}{I(0,\Lambda,\gamma)}$$
$$k_{\text{ox}}^{\text{MH}} = k_0 \frac{I(\theta,\Lambda,\gamma)}{I(0,\Lambda,\gamma)} \tag{4.33}$$

where Λ is the dimensionless reorganisation energy,

$$\Lambda = \lambda \frac{F}{RT} \tag{4.34}$$

and $I(\theta, \Lambda)$ is an integral over the energy levels of the metallic electrode of the form[6]

$$I(\theta, \Lambda, \gamma) = \int_{-\infty}^{\infty} \frac{\exp\left[\frac{-\Delta G_{\text{red/ox}}^{\ddagger}(\varepsilon)}{RT}\right]}{1 + \exp\left(\mp\varepsilon\right)} \, d\varepsilon \tag{4.35}$$

where ε is a dimensionless integral variable. The integral only needs to be calculated in the ε range where the integrand value is significant, typically ± 50, leading to very good accuracy. With this aim, the most straightforward numerical integration technique is the well-known trapezium rule (see Chapter 9).

---

[5] The symmetric version has been demonstrated to perform poorly for redox couples dissolved in solution [8].

[6] When a double sign appears, the upper sign refers to reduction and the lower one to oxidation.

The activation energy of the reduction/oxidation processes $(\Delta G^{\ddagger}_{\text{red/ox}}(\varepsilon))$ in the asymmetric version of the MH model is calculated from

$$
\frac{\Delta G^{\ddagger}_{\text{red/ox}}(\varepsilon)}{\mathcal{R}\mathcal{T}} = \frac{\Lambda}{4}\left(1 \pm \frac{\theta + \varepsilon}{\Lambda}\right)^2
$$

$$
+ \gamma \left(\frac{\theta + \varepsilon}{4}\right)\left\{1 - \left(\frac{\theta + \varepsilon}{\Lambda}\right)^2\right\} + \frac{\Lambda}{16}\gamma^2 \qquad (4.36)
$$

where γ is the dimensionless asymmetric parameter[7] that can take values in the range $-1 \leq \gamma \leq +1$ (although the model is not accurate for $|\gamma| > \frac{1}{3}$). Thus, a positive γ value is associated with the case where the vibrational force constants of the oxidised species are larger (in average) than the reduced one, and the opposite is true when a negative γ value is determined. The $|\gamma|$ value is directly related to the magnitude of these differences between vibrational force constants of oxidised and reduced species [9, 10]. Note that the well-known symmetric form of the Marcus–Hush model is a particular case of the asymmetric model for equal average force constants such that $\gamma = 0$.

According to all the above, in the aMH model the electrode kinetics are parameterised in function of four parameters: E_f^0, k_0, λ and γ. In spite of it having a more complex formulation than BV, only one additional parameter has been introduced and analytical expressions are available for the rate constants that enable simple implementation in simulations. Above all, the aMH model provides physical insight into the factors affecting the electrode kinetics, it enables predictions and offers suitable description of the voltammetry of surface-bound systems.

4.3. Unequal Diffusion Coefficients

For the more general case where the electroactive species A and B have different diffusion coefficients, we face the mathematical problem of two species according to the diffusion equations

$$
\frac{\partial c_A}{\partial t} = D_A \frac{\partial^2 c_A}{\partial x^2}
$$

$$
\frac{\partial c_B}{\partial t} = D_B \frac{\partial^2 c_B}{\partial x^2}
$$

(4.37)

---

[7] Not to be confused with the γ_i coefficients for the Thomas algorithm.

and the corresponding boundary conditions

$$\left. \begin{array}{l} t = 0, \ x \geq 0 \\ t > 0, \ x \to \infty \end{array} \right\} \quad c_A = c_A^*, \quad c_B = c_B^* \tag{4.38}$$

$$t > 0, \ x = 0: \quad D_A \left(\frac{\partial c_A(x,t)}{\partial x} \right)_{x=0} = k_{red} c_A(0,t) - k_{ox} c_B(0,t) \tag{4.39}$$

$$t > 0, \ x = 0: \quad D_B \left(\frac{\partial c_B(x,t)}{\partial x} \right)_{x=0} = - D_A \left(\frac{\partial c_A(x,t)}{\partial x} \right)_{x=0} \tag{4.40}$$

where Eq. (4.40) establishes that the flux of A reacting at the electrode is equal to the flux of B generated by conservation of mass.

Introducing the dimensionless variables $X = x/\epsilon$, $T = D_A t/\epsilon^2$, $d_B = D_B/D_A$ and $K_0 = k_0 \epsilon/D_A$ the differential equation problem becomes

$$\frac{\partial C_A}{\partial T} = \frac{\partial^2 C_A}{\partial X^2}$$
$$\frac{\partial C_B}{\partial T} = d_B \frac{\partial^2 C_B}{\partial X^2} \tag{4.41}$$

$$\left. \begin{array}{l} T = 0, \ X \geq 0 \\ T > 0, \ X \geq X_{max} \end{array} \right\} \quad C_A = 1, \quad C_B = C_B^* \tag{4.42}$$

$$T > 0, \ X = 0: \quad \left(\frac{\partial C_A(X,T)}{\partial X} \right)_{X=0} = f(\theta) K_0 \left[C_A(0,T) - C_B(0,T) e^{\theta} \right] \tag{4.43}$$

$$T > 0, \ X = 0: \quad d_B \left(\frac{\partial C_B(X,T)}{\partial X} \right)_{X=0} = - \left(\frac{\partial C_A(X,T)}{\partial X} \right)_{X=0} \tag{4.44}$$

where $C_B^* = c_B^*/c_A^*$ and $X_{max} = 6\sqrt{d_{max} T_{max}}$, with d_{max} being the largest dimensionless diffusion coefficient.

Following the procedure detailed in Chapter 3 for the case of unequal diffusion coefficients, with a non-uniform spatial grid, central difference scheme for the spatial derivative and the backward implicit time integration, the coefficients for the Thomas algorithm for species j = A, B are given by

$$\alpha_{i,j} = -d_j \left(\frac{2\Delta T}{\Delta X_-^2 + \Delta X_+ \Delta X_-} \right) \tag{4.45}$$

$$\beta_{i,j} = d_j \left(\frac{2\Delta T}{\Delta X_+^2 + \Delta X_+ \Delta X_-} + \frac{2\Delta T}{\Delta X_-^2 + \Delta X_+ \Delta X_-} \right) + 1 \tag{4.46}$$

$$\gamma_{i,j} = -d_j \left(\frac{2\Delta T}{\Delta X_+^2 + \Delta X_+ \Delta X_-} \right) \tag{4.47}$$

Using the two-point approximation for the surface gradients, the surface boundary conditions can be expressed as

$$T > 0, \; X = 0 : \quad \frac{C_{A,1}^k - C_{A,0}^k}{h} = f(\theta^{(k)})K_0 \left[C_{A,0}^k - C_{B,0}^k e^{\theta^{(k)}} \right] \tag{4.48}$$

$$T > 0, \; X = 0 : \quad d_B \left(C_{B,1}^k - C_{B,0}^k \right) = - \left(C_{A,1}^k - C_{A,0}^k \right) \tag{4.49}$$

that after discretisation take the form

$$C_{0,A}^k \left[1 + hf(\theta^{(k)})K_0 \right] - C_{0,B}^k hf(\theta^{(k)})K_0 e^{\theta^{(k)}} - C_{1,A}^k = 0$$

$$-C_{0,A}^k \frac{1}{d_B} hf(\theta^{(k)})K_0 + C_{0,B}^k \left[1 + \frac{1}{d_B} hf(\theta^{(k)})K_0 e^{\theta^{(k)}} \right] - C_{1,B}^k = 0 \tag{4.50}$$

Thus, the problem is formulated in terms of both species A and B. By ordering appropriately the equations and unknowns (concentrations of A and B), the system becomes tridiagonal and it can be solved with the Thomas algorithm. The left-hand side of the equation system expressed in matrix form is given by

$$
\begin{pmatrix}
\beta_{n-1,A} & \alpha_{n-1,A} & 0 & \cdots \\
\gamma_{n-2,A} & \beta_{n-2,A} & \alpha_{n-2,A} & 0 & \cdots \\
& \ddots & \ddots & \ddots \\
\cdots & 0 & \gamma_{1,A} & \beta_{1,A} & \alpha_{1,A} & 0 & \cdots \\
\cdots & 0 & 0 & \gamma_{0,A} & \beta_{0,A}^{(k)} & \alpha_{0,A}^{(k)} & 0 & \cdots \\
\cdots & 0 & 0 & 0 & \alpha_{0,B}^{(k)} & \beta_{0,B}^{(k)} & \gamma_{0,B} & 0 & \cdots \\
\cdots & 0 & 0 & 0 & 0 & \alpha_{1,B} & \beta_{1,B} & \gamma_{1,B} & 0 & \cdots \\
& & & & & \ddots & \ddots & \ddots \\
& & & & & \alpha_{n-2,B} & \beta_{n-2,B} & \gamma_{n-2,B} \\
& & & & & & \alpha_{n-1,B} & \beta_{n-1,B}
\end{pmatrix}
\begin{pmatrix}
C_{n-1,A} \\
C_{n-2,A}^k \\
\vdots \\
C_{1,A}^k \\
C_{0,A}^k \\
C_{0,B}^k \\
C_{1,B}^k \\
\vdots \\
C_{n-2,B}^k \\
C_{n-1,B}
\end{pmatrix}
$$

$$\tag{4.51}$$

where

$$\alpha_{0,A}^{(k)} = -hf(\theta^{(k)})K_0 e^{\theta^{(k)}}, \quad \beta_{0,A}^{(k)} = 1 + hf(\theta^{(k)})K_0, \quad \gamma_{0,A} = -1, \quad \delta_{0,A} = 0$$

$$\alpha_{0,B}^{(k)} = -\frac{1}{d_B}hf(\theta^{(k)})K_0, \quad \beta_{0,B}^{(k)} = 1 + \frac{1}{d_B}hf(\theta^{(k)})K_0 e^{\theta^{(k)}}, \quad \gamma_{0,B} = -1, \quad \delta_{0,B} = 0$$

$$(4.52)$$

Note that simulations with unequal diffusion coefficients can be tested with the criteria given in Subsubsection 3.5.2.2 where the case $d_B \neq 1$ was already considered. Thus, for reversible processes at the macroelectrodes the voltammograms shift towards more positive potentials as d_B increases according to Eq. (3.85) for CV and (3.91) for SWV.

4.4. Other One-Dimensional Electrode Geometries

Thus far we have considered only the case of planar macroelectrodes (i.e., linear diffusion conditions). Although these are widely used for electrochemical experiments, they have some drawbacks mainly due to the distorting effects arising from large capacitance and ohmic drop. In addition, mass transport in linear diffusion is quite inefficient such that in the case of fast homogeneous and heterogeneous reactions, the response is diffusion-limited and therefore it does not provide kinetic information.

In order to solve the above problems, miniaturisation of the working electrode has been common practice since the 1980s [11, 12]. Moreover, miniaturised electrodes provide steady-state responses and enable experimental studies in very reduced volumes. From a theoretical point of view, the simplest case corresponds to the use of spherical or hemispherical microelectrodes given that the system can be modelled simply in a one-dimensional coordinate system: r, the distance from the electrode centre (see Figure 4.4). Hence, the study of (hemi)spherical electrodes is of great interest to analyse the effect of convergent diffusion on the voltammetric response and introduce the concept of steady state. We will additionally consider the case of cylindrical electrodes as these can also be modelled in one spatial dimension provided that the length is sufficiently long such that the effect of the ends can be ignored.

Fick's second law for spherical electrodes in the one-dimensional space defined in Figure 4.4 is given by

$$\frac{\partial c}{\partial t} = D\left(\frac{\partial^2 c}{\partial r^2} + \frac{2}{r}\frac{\partial c}{\partial r}\right) \tag{4.53}$$

Cylindrical Electrode   **Spherical Electrode**

Fig. 4.4.   One-dimensional coordinate systems corresponding to cylindrical and spherical electrodes.

and for cylindrical electrodes by

$$\frac{\partial c}{\partial t} = D \left(\frac{\partial^2 c}{\partial r^2} + \frac{1}{r} \frac{\partial c}{\partial r} \right) \tag{4.54}$$

In both cases, we normalise the spatial coordinate against the radius of the electrode, r_e, such that the dimensionless distance, R, is given by

$$R = \frac{r}{r_e} \tag{4.55}$$

Note that the simulation space is now constrained to the region $1 \leq R \leq 1 + 6\sqrt{T_{max}}$; $R = 0$ is the point at the centre of the electrode (its interior) so naturally the simulation space begins at $R = 1$ ($r = r_e$), the electrode surface. The time is likewise normalised against the radius:

$$T = \frac{Dt}{r_e^2} \tag{4.56}$$

The diffusion equations can thus be generalised as

$$\frac{\partial C}{\partial T} = \frac{\partial^2 C}{\partial R^2} + \frac{\xi}{R} \frac{\partial C}{\partial R} \tag{4.57}$$

where $\xi = 1$ for a cylindrical geometry and 2 for a spherical geometry.

If we turn now to the discretisation of Fick's second law, the use of the second-order central difference approximation for the first derivative leads to

$$\frac{\partial C}{\partial R} = \frac{C_{i+1} - C_{i-1}}{\Delta R_+ + \Delta R_-} \tag{4.58}$$

where

$$\Delta R_+ + \Delta R_- = (R_{i+1} - R_i) + (R_i - R_{i-1}) = R_{i+1} - R_{i-1} \qquad (4.59)$$

Taking into account the above definitions and following the procedure detailed in Chapter 3, we obtain the following expressions for the α, β and γ coefficients for the Thomas algorithm of species j:

$$\alpha_i = d_j \left[-\frac{2\Delta T}{\Delta R_-^2 + \Delta R_+ \Delta R_-} + \frac{\xi}{R_i} \left(\frac{\Delta T}{\Delta R_+ + \Delta R_-} \right) \right] \qquad (4.60)$$

$$\beta_i = d_j \left[\frac{2\Delta T}{\Delta R_+^2 + \Delta R_+ \Delta R_-} + \frac{2\Delta T}{\Delta R_-^2 + \Delta R_+ \Delta R_-} \right] + 1 \qquad (4.61)$$

$$\gamma_i = d_j \left[-\frac{2\Delta T}{\Delta R_+^2 + \Delta R_+ \Delta R_-} - \frac{\xi}{R_i} \left(\frac{\Delta T}{\Delta R_+ + \Delta R_-} \right) \right] \qquad (4.62)$$

The surface conditions at $R = 1$ are equivalent to those given by Eqs. (4.48) and (4.49) with $K_0 = k_0 r_e / D$. Once the concentration profiles are known, the dimensionless current response is calculated from

$$J = \frac{I\,r_e}{FAD_A c_A^*} = -\left(\frac{\partial C_A}{\partial R} \right)_{R=1} \qquad (4.63)$$

where A is the area of the electrode ($= 4\pi r_e^2$ for spherical and $= 2\pi r_e^2$ for hemispherical electrodes).

4.4.1. *Testing Simulations: Specific tests*

As shown in Figure 4.5, at (hemi)spherical electrodes different regimes can be distinguished depending on the size of the electrode relative to the time scale of the experiment, that is, on the value of $T_{\max}$. In this section, several simple expressions will be given to test our simulations in all of these regimes.

4.4.1.1. *Small $T_{\max}$ limit: Macroelectrodes*

When the (hemi)spherical electrode is large and/or the time scale of the experiment is very short, the solution region affected by the electrochemical perturbation is very small in comparison with the electrode size such that the surface of the sphere would appear to be flat (as does the surface of the earth to the people living on it). Under these conditions, the diffusion is approximately linear (see Figure 4.5) and then the response is analogous to that described for planar macroelectrodes. Thus, all the criteria discussed in

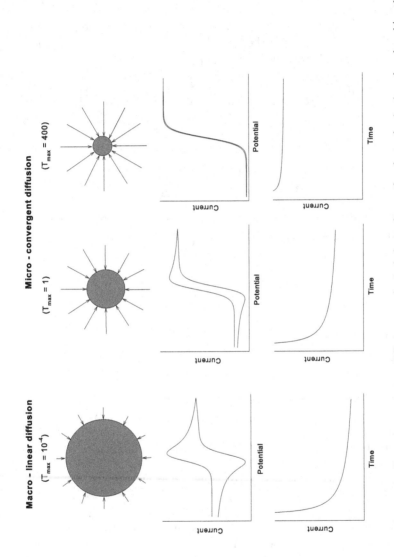

Fig. 4.5.   Transition from macrolinear to convergent diffusion at spherical electrodes as the electrode size is reduced and/or the time scale of the experiment is longer. Typical responses in cyclic voltammetry (middle) and chronoamperometry (bottom) are shown in each regime.

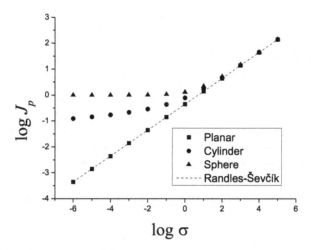

Fig. 4.6. The voltammetry peak height, J_{p}, as it varies with scan rate, σ, for planar, cylindrical and spherical electrodes for $K^0 = 10^{10}$.

Subsubsection 3.5.2.2 (for reversible electron transfers) and Subsection 4.2.2 (for non-reversible processes) are applicable. For example, in the limit of very fast scan rates, σ, independently of the geometry of the electrode the voltammetry will show a peaked response with the peak current being given by the Randles–Ševčík equation (see Figure 4.6).

4.4.1.2. *Intermediate $T_{\max}$ regime*

For smaller electrodes and/or longer times, the diffusion layer thickness relative to the electrode size is greater and the mass transport is more efficient given that the ratio between the volume of solution providing electroactive species and the electrode area is larger. Therefore, mass transport by diffusion is more effective and the decay of the surface gradient with time is reduced.

In this regime, the limiting current chronoamperograms (and so independently of the electron transfer kinetics and of the diffusion coefficient of the product of the electrode reaction, d_{B}) is given by

$$|J(T)| = 1 + \frac{1}{\sqrt{\pi T}} \tag{4.64}$$

and the concentration profile of the reactant (species A) by

$$C_A(R,T) = 1 - \frac{1}{R}\,\mathrm{erfc}\left(\frac{R-1}{2\sqrt{T}}\right) \tag{4.65}$$

4.4.1.3. *Large T_{max} limit: Steady state*

For spherical microelectrodes in the limit of slow scan rates $(T \gg 1)$, the diffusive mass transport is able to keep the surface gradient constant with time[8] and the steady state is attained. Under these conditions, a sigmoidal response is obtained in cyclic voltammetry with the current reaching a plateau rather than a peak as the voltage is increased (Figure 4.5). The cylinder will not show the same behaviour however; even though the radius of the cylinder may be of microscale, the length is of macroscale so diffusion to the cylinder is not totally convergent (though it does have a convergent component).

The stationary current response at spherical microelectrodes is not dependent on the scan rate, but on the applied potential according to the following expression:

$$J_{ss} = -\frac{K_{red}\left(1 - \frac{c_B^*}{c_A^*}e^\theta\right)}{1 + K_{red}\left(1 + \frac{e^\theta}{d_B}\right)} \qquad (4.66)$$

that in the limit of fully-reversible transfers turns into

$$J_{ss} = -\frac{1 - \frac{c_B^*}{c_A^*}e^\theta}{1 + \frac{e^\theta}{d_B}} \qquad (4.67)$$

with the reversible half-wave potential under these conditions being given by

$$\theta_{1/2,ss} = \ln(d_B) \qquad (4.68)$$

Consequently, for a given system, the reverse sweep of the voltammogram exactly overlaps the forward one so that the voltammogram appears to only sweep in one direction as shown in Figure 4.5.

The kinetics of a steady-state voltammogram can be extracted by examining the position of the half-wave potential, $\theta_{1/2}$, which is the potential at which the current reaches half of its maximum value. The variation of the half-wave potential with K^0 is shown in Figure 4.7. In the irreversible limit, the half-wave potential varies according to

$$\theta_{1/2} = \frac{\ln K^0}{\alpha} \qquad (4.69)$$

---

[8] We assume that the bulk solution is infinite in extent.

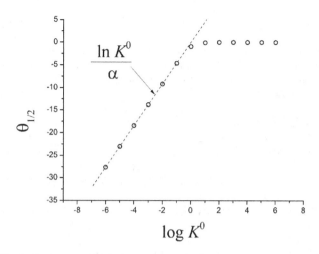

Fig. 4.7.   The half-wave potential, $\theta_{1/2}$, of the steady-state voltammetry of a spherical microelectrode as it varies with K^0 for $\sigma = 10^{-5}$.

In the SWV technique, the half-peak width and the peak position are not affected by the geometry of the electrode in the case of fully reversible electron transfers so that the expressions given for $W_{1/2}$ and θ_p in Subsection 3.5.2 are valid for (hemi)spherical electrodes. With respect to the peak current, this can be calculated under steady state conditions from the following simple expression [13]

$$|\Delta J_{\mathrm{p,ss}}| = \tanh\left(\frac{\theta_{\mathrm{SW}}}{2}\right) \qquad (4.70)$$

References

[1]   C. Amatore, O. Klymenko, and I. Svir, A new strategy for simulation of electrochemical mechanisms involving acute reaction fronts in solution: Principle, *Electrochem. Commun.* **12**, 1170–1173, (2010).

[2]   L. K. Bieniasz and C. Bureau, Use of dynamically adaptive grid techniques for the solution of electrochemical kinetic equations: Part 7. Testing of the finite-difference patch-adaptive strategy on example kinetic models with moving reaction fronts, in one-dimensional space geometry, *J. Electroanal. Chem.* **481**, 152–167, (2000).

[3]   M. Rudolph, Digital simulations on unequally spaced grids. Part 1. Critical remarks on using the point method by discretisation on a transformed grid, *J. Electroanal. Chem.* **529**, 97–108, (2002).

[4]   F. Martinez-Ortiz, A. Molina, and E. Laborda, Electrochemical digital simulation with high expanding four point discretization: Can Crank-Nicolson

uncouple diffusion and homogeneous chemical reactions?, *Electrochim. Acta.* **56**, 5707–5716, (2011).

[5] O. V. Klymenko, R. G. Evans, C. Hardacre, I. B. Svir, and R. G. Compton, Double potential step chronoamperometry at microdisk electrodes: Simulating the case of unequal diffusion coefficients, *J. Electroanal. Chem.* **571**, 211–221, (2004).

[6] D. Britz and J. Strutwolf, *Digital Simulation in Electrochemistry. 4th ed.* (Springer, Berlin, 2016).

[7] C. E. D. Chidsey, Free energy and temperature dependence of electron transfer at the metal-electrolyte interface, *Science.* **215**, 919–922, (1991).

[8] M. C. Henstridge, E. Laborda, N. V. Rees, and R. G. Compton, Marcus-Hush-Chidsey theory of electron transfer applied to voltammetry: A review, *Electrochim. Acta.* **84**, 12–20, (2012).

[9] R. A. Marcus, On the theory of electron-transfer reactions. VI. Unified treatment for homogeneous and electrode reactions, *J. Chem. Phys.* **43**, 679–701, (1965).

[10] E. Laborda, M. C. Henstridge, and R. G. Compton, Asymmetric Marcus theory: Application to electrode kinetics, *J. Electroanal. Chem.* **667**, 48–53, (2012).

[11] M. I. Montenegro, M. A. Queiros, and J. L. Daschbach, *Microelectrodes: Theory and Applications.* vol. 197, (NATO ASI Series E, 1990).

[12] K. J. Aoki and J. Chen, Insight of electrolyte-free voltammetry at microelectrodes, *Current Opinion in Electrochemistry.* **10**, 67–71, (2018).

[13] A. Molina and J. González, *Pulse Voltammetry in Physical Electrochemistry and Electroanalysis.* (Springer, Berlin, 2016).

Chapter 5

First-Order Chemical Kinetic Mechanisms

So far we have considered the so-called E mechanism where a simple one-electron transfer is the only process that takes place within the time scale of the experiments. However, in many situations the electrochemical system is more complex and includes more than one electron transfer and/or coupled chemical reactions, as has been discussed in Chapter 1 [1–3]. These processes affect the concentration of the electroactive species and thus the electrochemical response, which is the basis of the use of such techniques for the characterisation of reaction mechanisms and kinetics.

From a theoretical point of view, the existence of coupled processes complicates the numerical resolution of the problem. Depending on the reaction mechanism, these complications may affect simply the form of the coefficients of the Thomas algorithm (as in the first-order EC_{irre} mechanism, see Section 5.1) or, more profoundly, the problem-solving algorithm due to coupled (e.g., first-order catalytic and EC_{rev} mechanisms, see Sections 5.2 and 5.4) and non-linear (see Chapter 6) equation systems.

In this chapter, the case of first-order mechanisms will be considered, analysing the changes in the corresponding differential equations, boundary conditions and problem-solving methodology. Strategies for the resolution of the linear equation systems resulting from discretisation will be detailed for some representative mechanisms. The case of multi-electron transfers will also be considered. This will enable us to simulate other common situations where the electroactive molecule transfers more than one electron. For the sake of simplicity, linear diffusion conditions (macroelectrodes) are considered, although the treatment is analogous for any other electrode geometry by modifying the terms corresponding to the diffusion mass transport as discussed in Section 4.4 and Chapters 9 and 10.

5.1. First-Order EC$_{\text{irre}}$ Mechanism: Basic Concepts

The first reaction mechanism we consider is the EC$_{\text{irre}}$ mechanism, where E refers to a heterogeneous electron transfer reaction and C$_{\text{irre}}$ to an irreversible chemical reaction in solution. The EC$_{\text{irre}}$ case enables us to introduce some basic concepts in the resolution of diffusive-kinetic electrochemical problems without including significant additional complications in the numerical procedure. The first-order EC$_{\text{irre}}$ scheme is defined by

$$A + e^- \rightleftarrows B$$
$$B \xrightarrow{k_1} Y \tag{5.1}$$

where $k_1(s^{-1})$ is the first-order homogeneous rate constant and species Y is assumed to be electroinactive in the potential region of study. Experimentally this situation can correspond to the decomposition of species B or to its reaction with a species that is present in a large excess such that its concentration can be considered as constant (pseudo-first-order conditions), such as in reactions with the solvent.

Considering scheme (5.1), the problem includes three unknowns: the concentrations of species A, B and Y that each depends on both time and distance to the electrode surface. The concentration of species A in solution is only subject to transport by diffusion such that its variation with time and distance is described by Fick's second law; thus, for linear diffusion

$$\frac{\partial c_A}{\partial t} = D_A \frac{\partial^2 c_A}{\partial x^2} \tag{5.2}$$

Note that this does not mean that the concentration profile of species A is equivalent to that of the E mechanism since it will be influenced by the chemical reaction through the surface boundary conditions.[1] Thus, the chemical reaction affects the surface concentration of species B, which is linked to that of species A through the Nernst equation (for reversible systems), or more generally, through the Butler–Volmer or Marcus–Hush relationships. Therefore, the surface concentration of species A, and as a consequence the whole concentration profile, will reflect the presence of the chemical process.

---

[1] Under diffusion-limited conditions the concentration profile of species A is unaffected by the chemical reaction involving species B, given that in this situation the surface condition establishes that $c_A(0) = 0$, which is independent of species B and its reactivity.

Regarding species B and Y, their concentrations will undertake diffusional and chemical changes simultaneously. Accordingly, an additional term associated with the conversion of species B into Y appears in the continuity equations

$$\frac{\partial c_B}{\partial t} = D_B \frac{\partial^2 c_B}{\partial x^2} - k_1 c_B \tag{5.3}$$

$$\frac{\partial c_Y}{\partial t} = D_Y \frac{\partial^2 c_Y}{\partial x^2} + k_1 c_B \tag{5.4}$$

The corresponding boundary value problem includes zero flux at the electrode surface for species Y given that it is supposed to be electroinactive in the potential range of the experiments. We assume that species A is the only one initially present in solution:

$$\left. \begin{array}{l} t = 0, \ x \geq 0 \\ t > 0, \ x \to \infty \end{array} \right\} \quad c_A = c_A^*, \quad c_B = 0, \quad c_Y = 0 \tag{5.5}$$

$$t > 0, \ x = 0 \quad \left\{ \begin{array}{l} D_A \left(\dfrac{\partial c_A}{\partial x} \right)_{x=0} = k_{red} c_A(0) - k_{ox} c_B(0) \\[2ex] D_B \left(\dfrac{\partial c_B}{\partial x} \right)_{x=0} = - D_A \left(\dfrac{\partial c_A}{\partial x} \right)_{x=0} \\[2ex] \left(\dfrac{\partial c_Y}{\partial x} \right)_{x=0} = 0 \end{array} \right. \tag{5.6}$$

Considering the dimensionless parameters we have introduced in previous chapters ($X = x/\epsilon$, $T = D_A t/\epsilon^2$, $\theta = F(E - E_f^0)/\mathcal{R}\mathcal{T}$, $K_0 = k_0 \epsilon / D_A$, $d_j = D_j/D_A$ and $C_j = c_j/c_A^*$), the differential equation system and boundary value problem can be rewritten as

$$\frac{\partial C_A}{\partial T} = \frac{\partial^2 C_A}{\partial X^2}$$

$$\frac{\partial C_B}{\partial T} = d_B \frac{\partial^2 C_B}{\partial X^2} - K_1 C_B \tag{5.7}$$

$$\frac{\partial C_Y}{\partial T} = d_Y \frac{\partial^2 C_Y}{\partial X^2} + K_1 C_B$$

$$\left. \begin{array}{l} T = 0, \ X \geq 0 \\ T > 0, \ X \to \infty \end{array} \right\} \quad C_A = 1, \quad C_B = 0, \quad C_Y = 0 \tag{5.8}$$

$$T > 0, \ X = 0 \quad \begin{cases} \left(\dfrac{\partial C_A}{\partial X}\right)_{X=0} = f(\theta) K_0 \left[C_A(0) - C_B(0)e^{\theta}\right] \\[2ex] d_B \left(\dfrac{\partial C_B}{\partial X}\right)_{X=0} = -\left(\dfrac{\partial C_A}{\partial X}\right)_{X=0} \\[2ex] \left(\dfrac{\partial C_Y}{\partial X}\right)_{X=0} = 0 \end{cases} \quad (5.9)$$

where the dimensionless chemical rate constant, K_1, is defined as

$$K_1 = \frac{k_1 \epsilon^2}{D_A} \quad (5.10)$$

By approximating the spatial derivatives according to the central three-point difference formula and using the backward implicit scheme, the form of the resulting equations for species A and B are analogous to those discussed in previous chapters:

$$\alpha_i C_{i-1}^k + \beta_i C_i^k + \gamma_i C_{i+1}^k = \delta_i^k \quad (5.11)$$

The only difference with respect to the E mechanism is that the coefficient β of the Thomas algorithm for species B has an additional term associated with the kinetics of the chemical reaction:

$$\beta_{i,B} = d_B \left(\frac{2\Delta T}{\Delta X_+^2 + \Delta X_+ \Delta X_-} + \frac{2\Delta T}{\Delta X_-^2 + \Delta X_+ \Delta X_-}\right) + 1 + \Delta T K_1 \quad (5.12)$$

Thus, the concentration profiles of species A and B can be obtained by applying the Thomas algorithm as discussed in Chapters 3 and 4. Finally, the current response corresponds to the interconversion of species A and B at the electrode surface and is given by

$$J = -\left(\frac{\partial C_A}{\partial X}\right)_{X=0} = -f(\theta) K_0 [C_A(0) - e^{\theta} C_B(0)] \quad (5.13)$$

Equation (5.12) illustrates how the concentration of species B changes with time not only due to mass transport by diffusion but also due to the homogeneous chemical reaction, with the value $\Delta T K_1$ providing an estimation of the variation of concentration in each timestep (ΔT) due to the chemical conversion of species B into Y.

In the reaction scheme (5.1) we have considered that the chemical process is irreversible such that the concentration profiles of the electroactive species and the electrochemical response are independent of the behaviour of species Y. Nevertheless, if we were interested in this, the concentration profile of Y could be calculated from that of species B:

$$\alpha_{i,\text{Y}} C^k_{i-1,\text{Y}} + \beta_{i,\text{Y}} C^k_{i,\text{Y}} + \gamma_{i,\text{Y}} C^k_{i+1,\text{Y}} = \delta^k_{i,\text{Y}} \qquad (5.14)$$

where the coefficients α, β and γ are equivalent to those presented in previous chapters and the coefficient δ^k now contains the term $K_1 \Delta T C^k_{i,\text{B}}$. This term is known since the profile of species B has been already obtained:

$$\delta^k_{i,\text{Y}} = C^{k-1}_{i,\text{Y}} + K_1 \Delta T C^k_{i,\text{B}} \qquad (5.15)$$

Note that the zero-flux surface condition for this species, after discretisation with the two-point approximation, establishes that

$$\frac{C^k_{1,\text{Y}} - C^k_{0,\text{Y}}}{h} = 0 \qquad (5.16)$$

such that the coefficients for the Thomas algorithm corresponding to the point at the electrode surface are given by

$$\alpha_{0,\text{Y}} = 0, \quad \beta_{0,\text{Y}} = -1, \quad \gamma_{0,\text{Y}} = 1, \quad \delta_{0,\text{Y}} = 0 \qquad (5.17)$$

where h is the first spatial interval: $h = X_1 - X_0$.

The effects of the chemical reaction on the cyclic and square wave voltammograms of the $E_{\text{rev}} C_{\text{irre}}$ mechanism at a macroelectrode are shown in Figure 5.1. For slow chemical reactions, the signals are only slightly affected by the presence of the coupled reaction whereas there is a very significant influence on the CV peak of the reverse scan corresponding to the electrooxidation of species B. As expected, because of the conversion of B into the electroinactive species Y, species B is consumed in the chemical process and the anodic peak decreases. For fast reactions, the CV backward peak fully disappears and there is a slight increase in the forward peak current (*ca.* 10%) with respect to that corresponding to a reversible E mechanism. With regard to SWV, initially the peak current decreases as the chemical reaction is faster and then it becomes independent of K_1. In both techniques, the electro-reduction signal shifts towards more positive potentials when K_1 increases.

5.1.1. *Reaction layer*

The inference of homogeneous coupled chemical reactions can require the refinement of the spatial grid to reflect the effect of the chemical process on

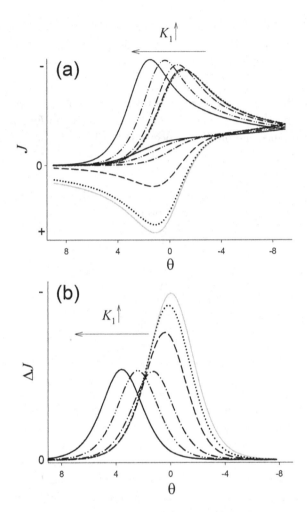

Fig. 5.1. Effect of the chemical kinetics on the (a) cyclic voltammetry and (b) square wave voltammetry of the first-order $E_{rev}C_{irre}$ mechanism at a macroelectrode. The grey line corresponds to the response of a simple electron transfer without coupled chemical reactions.

the concentration profiles. For the case of the first-order EC_{irre} this effect is shown in Figure 5.2 where the concentration profiles[2] of species A, B and Y corresponding to the peak of the cyclic voltammogram are plotted for different rate constants of the chemical reaction, K_1.

---

[2] The value $X/\sqrt{T_{max}}$ is chosen as x-axis since this gives an estimation of the distance relative to the linear diffusion layer thickness of the species: $X/\sqrt{T_{max}} = x/\sqrt{D_A t_{max}}$.

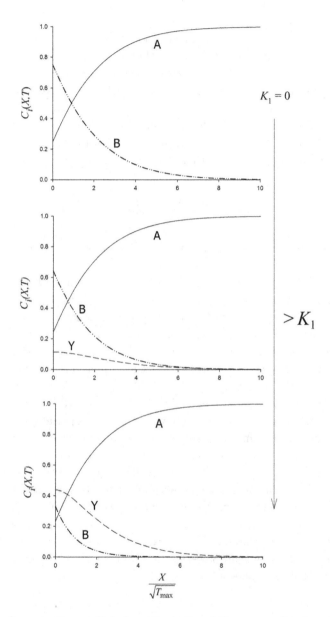

Fig. 5.2. Concentration profiles of species A, B and Y corresponding to a first-order $E_{rev}C_{irre}$ mechanism at a macroelectrode as the chemical kinetics is faster, that is, as K_1 increases.

When the chemical reaction is slow, the changes in concentration take place in a region of thickness similar to the diffusion-only problem analysed in the previous chapters of this book. Therefore, the spatial grid can be equivalent. The linear diffusion layer thickness, $\delta_{d,lin}$, gives an idea of the spatial region next to the electrode surface where changes of concentration are more important such that the effect of the spatial discretisation can be analysed in terms of the number of points included in the region $0 \leq x \leq \delta_{d,lin}$. The maximum $\delta_{d,lin}$ value is attained under diffusion-limited conditions [4] and for (hemi)spherical electrodes of radius r_e it is given by

$$\delta_{d,lin} = \left(\frac{1}{r_e} + \frac{1}{\sqrt{\pi D t}} \right)^{-1} \tag{5.18}$$

The above expression reproduces the results for planar electrodes by making $r_e \to \infty$.

As the chemical conversion of species B into Y is more rapid, the concentration profile of species B is compressed towards the electrode surface so that the most significant changes of species B concentration happen in a smaller region. Therefore, more points of the spatial grid must be placed in this region with respect to the grid for diffusion-only systems. Note that this means a serious limitation if explicit methods are used for simulations given the restrictions in the $\Delta T / \Delta X^2$ value discussed in Chapter 3. Unconditionally stable methods, like in the backward implicit approach followed in this book, in combination with expanding spatial grids (Chapter 4) enable us to overcome this problem and develop efficient simulations.

Within the above context, the concept of *reaction layer* is helpful [5] in order to provide an estimation of the solution region where the spatial grid must be refined. This is defined as the zone where the chemical equilibrium breaks down, that is, where the concentrations of the species are not in equilibrium. The thickness of the *linear* reaction layer, $\delta_{r,lin}$, for first-order reactions with k_1 (s^{-1}) and k_{-1} (s^{-1}) being the forward and backward rate constants, respectively, can be estimated from the following expression for (hemi)spherical electrodes [6]:

$$\delta_{reac} = \left(\frac{1}{r_e} + \sqrt{\frac{k_1 + k_{-1}}{D}} \right)^{-1} \tag{5.19}$$

The value for planar electrodes can again be deduced easily by making $r_e \to \infty$: $\delta_{reac}^{P} = \sqrt{\frac{D}{k_1 + k_{-1}}}$.

5.2. First-Order Catalytic Mechanism: Coupled Equation Systems

As discussed in the previous section, the EC_{irre} problem can be solved in a similar way to a simple charge transfer by making use of the Thomas algorithm. Nevertheless, when coupled chemical processes occur this generally leads to the coupling of the chemical-diffusion equation system, which prevents the use of the classical Thomas algorithm. In this section we consider one of these situations corresponding to the catalytic mechanism where the reactant A is regenerated by a chemical reaction of species B in solution and the redox species show unequal diffusion coefficients[3]

$$A + e^- \rightleftharpoons B$$

$$B + Y \underset{k_{-1}}{\overset{k_1}{\rightleftharpoons}} A + Z \tag{5.20}$$

If we assume that species Y and Z are in large excess ($c_Y^*, c_Z^* \gg c_A^*, c_B^*$), the chemical reactions are pseudo-first-order with the rate constants $k_1'(s^{-1}) = k_1 c_Y^*$ and $k_{-1}'(s^{-1}) = k_{-1}c_Z^*$. Thus, the differential equation system that describes the variation of the concentrations of species A and B under linear diffusion conditions is given by

$$\frac{\partial c_A}{\partial t} = D_A \frac{\partial^2 c_A}{\partial x^2} - k_{-1}' c_A + k_1' c_B$$

$$\frac{\partial c_B}{\partial t} = D_B \frac{\partial^2 c_B}{\partial x^2} + k_{-1}' c_A - k_1' c_B \tag{5.21}$$

and the boundary value problem by

$$\left. \begin{array}{l} t = 0, \ x \geq 0 \\ t > 0, \ x \to \infty \end{array} \right\} \quad c_A = c_A^*, \quad c_B = c_B^* \tag{5.22}$$

$$t > 0, \ x = 0 \quad \begin{cases} D_A \left(\dfrac{\partial c_A}{\partial x} \right)_{x=0} = k_{red} c_A(0) - k_{ox} c_B(0) \\[3mm] D_B \left(\dfrac{\partial c_B}{\partial x} \right)_{x=0} = - D_A \left(\dfrac{\partial c_A}{\partial x} \right)_{x=0} \end{cases} \tag{5.23}$$

Note that in this case both A and B are present in solution at the beginning of the experiment in the concentrations established by the chemical equilibrium:

---

[3] Note that in the case $d_B = 1$ the problem can be reduced to that of a single species noting that $c_A(x,t) + c_B(x,t) = c_A^* + c_B^*$, as discussed in Subsection 3.5.1.

$$K_{\text{eq}} = \frac{k'_1}{k'_{-1}} = \frac{c^*_A}{c^*_B} \qquad (5.24)$$

In practice it is necessary to calculate the equilibrium concentrations of the different species from those initially added to the working solution, c_{tot}. In the present case, assuming that a long enough equilibration time elapses before the measurement, the initial (equilibrium) concentrations of the reactant species A and the product B can be calculated from the value of the equilibrium constant (5.24) and considering that $c_{\text{tot}} = c^*_A + c^*_B$ such that

$$c^*_A = c_{\text{tot}} \left(\frac{K_{\text{eq}}}{1 + K_{\text{eq}}} \right) \qquad (5.25)$$

$$c^*_B = c_{\text{tot}} \left(\frac{1}{1 + K_{\text{eq}}} \right) \qquad (5.26)$$

After normalisation and discretisation of the problem with the central three-point approximation for the spatial derivative and the backward implicit scheme the differential equations of a point i in solution can be written in the form

$$\alpha_{i,A} C^k_{i-1,A} + \left(\beta^{\text{dif}}_{i,A} + K_{-1}\Delta T \right) C^k_{i,A} + \gamma_{i,A} C^k_{i+1,A} - K_1 \Delta T C^k_{i,B} = \delta^k_{i,A}$$

$$\alpha_{i,B} C^k_{i-1,B} + \left(\beta^{\text{dif}}_{i,B} + K_1 \Delta T \right) C^k_{i,B} + \gamma_{i,B} C^k_{i+1,B} - K_{-1} \Delta T C^k_{i,A} = \delta^k_{i,B}$$
$$(5.27)$$

where the form of the coefficients α, β^{dif} and γ is that given in previous chapters for diffusion-only systems and the values of $\delta^k_i = C^{k-1}_i$ are known.

As can be seen in (5.27), the equations to solve have terms of more than one species, leading to a coupled, linear equation system. This complicates the resolution of the problem given that the Thomas algorithm cannot be applied in its classical form (see below). Approximate methods have been proposed in the literature to tackle this situation by forcing the separation of the diffusional and chemical changes of concentrations [7 9]. A more general and accurate procedure will be considered next.

By appropriate ordering of the unknowns,

$$\left[C^k_{0,A} \, C^k_{0,B} \, C^k_{1,A} \, C^k_{1,B} \, \cdots \, C^k_{43,A} \, C^k_{43,B} \, \cdots \, C^k_{n-2,A} \, C^k_{n-2,B} \, C^k_{n-1,A} \, C^k_{n-1,B} \right]^T$$
$$(5.28)$$

a *pentadiagonal* system of linear equations is obtained:

$$\mathbf{Ax} = \mathbf{b} \qquad (5.29)$$

where the left-hand-side term is given by

$$
\begin{pmatrix}
\beta_{0,A}^{(k)} & -hK_{ox}^{(k)} & \gamma_{0,A} & 0 & 0 & 0 & 0 & 0 & 0 & 0 & 0 & 0 & \cdots & & & & & \cdots \\
-\tfrac{1}{d_B}hK_{red}^{(k)} & \beta_{0,B}^{(k)} & 0 & \gamma_{0,B} & 0 & 0 & 0 & 0 & 0 & 0 & 0 & 0 & \cdots & & & & & \cdots \\
\alpha_{1,A} & 0 & \beta_{1,A} & -K_1^* & \gamma_{1,A} & 0 & 0 & 0 & 0 & 0 & 0 & 0 & \cdots & & & & & \cdots \\
0 & \alpha_{1,B} & -K_{-1}^* & \beta_{1,B} & 0 & \gamma_{1,B} & 0 & 0 & 0 & 0 & 0 & 0 & & & & & & \\
0 & 0 & \alpha_{2,A} & 0 & \beta_{2,A} & -K_1^* & \gamma_{2,A} & 0 & 0 & 0 & 0 & 0 & & & & & & \\
0 & 0 & 0 & \alpha_{2,B} & -K_{-1}^* & \beta_{2,B} & 0 & \gamma_{2,B} & 0 & 0 & 0 & 0 & & & & & & \\
& & & & \ddots & \ddots & \ddots & \ddots & & & & & & & & & & \\
& & & & & & \alpha_{43,A} & 0 & \beta_{43,A} & -K_1^* & \gamma_{43,A} & 0 & & & & & & \\
& & & & & & 0 & \alpha_{43,B} & -K_{-1}^* & \beta_{43,B} & 0 & \gamma_{43,B} & & & & & & \\
& & & & & & 0 & 0 & \alpha_{44,A} & 0 & \beta_{44,A} & -K_1^* & \gamma_{44,A} & & & & & \\
& & & & & & 0 & 0 & 0 & \alpha_{44,B} & -K_{-1}^* & \beta_{44,B} & \gamma_{44,B} & & & & & \\
& & & & & & & & & & \ddots & \ddots & \ddots & \ddots & & & & \\
\vdots & & & & & & & & & & & & & \alpha_{n-2,A} & 0 & \beta_{n-2,A} & -K_1^* & \gamma_{n-2,A} & 0 \\
\vdots & & & & & & & & & & & & & 0 & \alpha_{n-2,B} & -K_{-1}^* & \beta_{n-2,B} & 0 & \gamma_{n-2,B} \\
\vdots & & & & & & & & & & & & & 0 & 0 & \alpha_{n-1,A} & 0 & \beta_{n-1,A} & 0 \\
\vdots & & & & & & & & & & & & & 0 & 0 & 0 & \alpha_{n-1,B} & 0 & \beta_{n-1,B}
\end{pmatrix}
\tag{5.30}
$$

and the right-hand-side term by

$$\left[\,\delta_{0,A}\;\delta_{0,B}\;\delta_{1,A}^k\;\delta_{1,B}^k\;\cdots\;\delta_{43,A}^k\;\delta_{43,B}^k\;\cdots\;\delta_{n-2,A}^k\;\delta_{n-2,B}^k\;\delta_{n-1,A}\;\delta_{n-1,B}\,\right]^T$$
$$\text{(5.31)}$$

with $K^*_{1/-1} = K_{1/-1}\Delta T$ and n being the number of points of the spatial grid.

According to Eq. (5.27) the β coefficients in the coefficient matrix $\mathbf{A}$ include a new term associated with the chemical process

$$\beta_{i,A} = \beta_{i,A}^{\text{dif}} + K_{-1}\Delta T$$
$$\beta_{i,B} = \beta_{i,B}^{\text{dif}} + K_1\Delta T$$
$$\text{(5.32)}$$

and the first equations of the system correspond to the surface conditions as discussed in Chapter 4,

$$C_{0,A}^k\left[1 + hf(\theta^{(k)})K_0\right] - C_{0,B}^k hf(\theta^{(k)})K_0 e^{\theta^{(k)}} - C_{1,A}^k = 0$$
$$-C_{0,A}^k\frac{1}{d_B}hf(\theta^{(k)})K_0 + C_{0,B}^k\left[1 + \frac{1}{d_B}hf(\theta^{(k)})K_0 e^{\theta^{(k)}}\right] - C_{1,B}^k = 0$$
$$\text{(5.33)}$$

such that the coefficients of the first two rows in Eq. (5.30) are

$$\beta_{0,A}^{(k)} = 1 + hK_{\text{red}}^{(k)}, \quad \gamma_{0,A} = -1, \quad \delta_{0,A} = 0$$
$$\beta_{0,B}^{(k)} = 1 + \frac{1}{d_B}hK_{\text{ox}}^{(k)}, \quad \gamma_{0,B} = -1, \quad \delta_{0,B} = 0$$
$$\text{(5.34)}$$

with $K_{\text{red}} = f(\theta)K_0$ and $K_{\text{ox}} = f(\theta)K_0 e^{\theta}$.

As discussed in Chapter 3, the bulk condition for species j establishes the following coefficients for the point $n - 1$:

$$\alpha_{n-1,j} = 0, \quad \beta_{n-1,j} = 1, \quad \delta_{n-1,j} = \frac{c_j^*}{c_A^*} \tag{5.35}$$

For the resolution of the above problem a general algorithm for banded linear equation systems is detailed in Section 5.3. This enables us to design programs that cover a great variety of electrochemical processes (involving first-order kinetics) in a very general way, including the cases that we have studied previously, that is, the E and EC$_{\text{irre}}$ mechanisms. Although for the latter more efficient simulations can be performed noting their particular characteristics discussed previously, with modern computers the advantage in terms of simulation time is usually not relevant for typical electrochemical experiments.

Once the concentration profiles have been calculated, the current is obtained from Eq. (5.13). Figure 5.3 shows the variation of the cyclic and

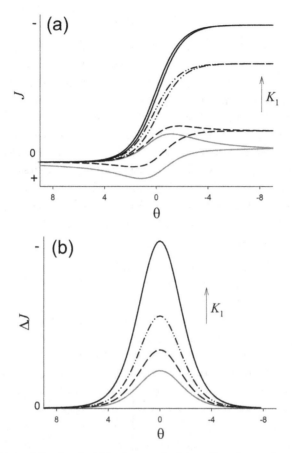

Fig. 5.3. Effect of the chemical kinetics on the (a) cyclic voltammetry and (b) square wave voltammetry of the first-order catalytic mechanism at a macroelectrode with a reversible electron transfer, equal diffusion coefficients and $K_{-1} = 0$. The grey line corresponds to the response of a simple electron transfer without coupled chemical reactions.

square wave voltammograms corresponding to a first-order catalytic mechanism where $k_{-1} = 0$, $d_B = 1$ and the electrode reaction is fully reversible. In CV, as the chemical reaction is faster and so is the regeneration of species A, the reduction current increases and the oxidation peak in the reverse scan disappears. Eventually, a sigmoidal, steady-state response is obtained where the responses of the forward and reverse scans overlap. In SWV, a peak-shaped response of the same width and position is obtained independently of the chemical rate constants, the peak height increasing with the K_1-value.

5.3. LU Decomposition and Extended Thomas Algorithm

When the coefficient matrix, $\mathbf{A}$, of the linear algebraic system $\mathbf{Ax} = \mathbf{b}$ is more than tridiagonal (such as in the problem of the first-order catalytic mechanism) then the Thomas algorithm in its classical form cannot be applied. With this aim, linear equation solvers (as well as other useful tools for our simulations) are available, for example, for C++ in the section "Linear Algebra" of the GNU Scientific Library (GSL).[4] Among the direct methods for solving systems of linear equations, next we consider the basics for the resolution through the LU decomposition of $\mathbf{A}$ and its application to diagonal systems that are frequently found in electrochemical simulation.

Let us consider the linear system $\mathbf{Ax} = \mathbf{LUx} = \mathbf{b}$ where the $N \times N$ matrix $\mathbf{A}$ can be factorised into a lower triangular matrix $\mathbf{L}$ and an upper triangular matrix $\mathbf{U}$ as

$$\mathbf{A} = \begin{pmatrix} a_{00} & a_{01} & a_{02} & \cdots & a_{0(N-1)} \\ a_{10} & a_{11} & a_{12} & \cdots & a_{1(N-1)} \\ a_{20} & a_{21} & a_{22} & \cdots & a_{2(N-1)} \\ \vdots & \vdots & & \ddots & \\ a_{(N-1)0} & a_{(N-1)1} & a_{(N-1)2} & \cdots & a_{(N-1)(N-1)} \end{pmatrix}$$

$$= \begin{pmatrix} 1 & 0 & 0 & \cdots & 0 \\ l_{10} & 1 & 0 & \cdots & 0 \\ l_{20} & l_{21} & 1 & \cdots & 0 \\ \vdots & \vdots & & \ddots & \\ l_{(N-1)0} & l_{(N-1)1} & l_{(N-1)2} & \cdots & 1 \end{pmatrix} \begin{pmatrix} u_{00} & u_{01} & u_{02} & \cdots & u_{0(N-1)} \\ 0 & u_{11} & u_{12} & \cdots & u_{1(N-1)} \\ 0 & 0 & u_{22} & \cdots & u_{2(N-1)} \\ \vdots & \vdots & & \ddots & \vdots \\ 0 & 0 & 0 & \cdots & u_{(N-1)(N-1)} \end{pmatrix}$$

$$(5.36)$$

where the coefficients l_{ij} and u_{ij} are given by

for $j = 0, 1, \ldots, N-1$

$$u_{ij} = a_{ij} - \sum_{k=0}^{i-1} l_{ik} u_{kj}, \quad i = 0, 1, \ldots, j$$

$$l_{ij} = \frac{1}{u_{jj}} \left(a_{ij} - \sum_{k=0}^{j-1} l_{ik} u_{kj} \right), \quad i = j+1, \; j+2, \ldots, N-1$$

$$(5.37)$$

---

[4] http://www.gnu.org/software/gsl.

Note that $l_{ii} = 1$ (for $i = 0, 1, \ldots, N - 1$) and the summations in all the expressions of this section are zero when the lower limit is larger than the upper one. Therefore we can write

$$\mathbf{Ax} = (\mathbf{LU}) \mathbf{x} = \mathbf{L} (\mathbf{Ux}) = \mathbf{b} \qquad (5.38)$$

and

$$\mathbf{Ly} = \mathbf{b} \qquad (5.39)$$

The solution of the above triangular set of equations can be performed easily by forward substitution:

$$y_i = b_i - \sum_{k=0}^{i-1} l_{ik} y_k, \quad i = 0, 1, \ldots, N - 1 \qquad (5.40)$$

Finally, the vector $\mathbf{x}$ can be determined by solving $\mathbf{Ux} = \mathbf{y}$ through backward substitution:

$$x_i = \frac{1}{u_{ii}} \left(y_i - \sum_{k=i+1}^{N-1} u_{ik} x_k \right), \quad i = N - 1, N - 2, \ldots, 0 \qquad (5.41)$$

The equation systems commonly found in the simulation of electrochemical problems have a diagonal structure with a bandwidth $2n + 1$ such that $a_{ij} = 0$ for $j > i + n$ and for $i > j + n$. By particularising the LU decomposition for diagonal systems (*extended Thomas algorithm*) the above expressions turn into

$j = 0, 1, \ldots, N - 1$:

$$
\begin{cases}
u_{ij} = a_{ij} - \displaystyle\sum_{k=\max(0,j-n)}^{i-1} l_{ik} u_{kj}, & i = \max (0, j - n), \ldots, j \\[4mm]
l_{ij} = \dfrac{1}{u_{jj}} \left(a_{ij} - \displaystyle\sum_{k=\max(0,i-n)}^{j-1} l_{ik} u_{kj} \right), & i = j + 1, \ldots, j + n \\[4mm]
l_{ii} = 1
\end{cases}
$$

$$y_i = b_i - \sum_{k=\max(0,i-n)}^{i-1} l_{ik} y_k, \quad i = 0, 1, \ldots, N - 1$$

$$x_i = \frac{1}{u_{ii}} \left(y_i - \sum_{k=i+1}^{\min(N-1,i+n)} u_{ik} x_k \right), \quad i = N - 1, N - 2, \ldots, 0$$

$$(5.42)$$

where $\max(a, b)$ is equal to the value of the largest argument, a or b, and $\min(a, b)$ to the smallest one.

5.4. First-Order EC_{rev} Mechanism: Including a Third Species

Next, the first-order EC_{rev} mechanism will be discussed as an example of electrochemical problems with a coupled equation system and more than two species.[5] The reaction scheme corresponding to the EC_{rev} mechanism can be written as

$$A + e^- \rightleftarrows B$$

$$B \underset{k_{-1}}{\overset{k_1}{\rightleftarrows}} Y \tag{5.43}$$

where species Y is electroinactive in the potential region considered. The differential equation system describing the mass transport under linear diffusion and concentration changes due to the chemical reaction is given by

$$\frac{\partial c_A}{\partial t} = D_A \frac{\partial^2 c_A}{\partial x^2}$$

$$\frac{\partial c_B}{\partial t} = D_B \frac{\partial^2 c_B}{\partial x^2} - k_1 c_B + k_{-1} c_Y \tag{5.44}$$

$$\frac{\partial c_Y}{\partial t} = D_Y \frac{\partial^2 c_Y}{\partial x^2} + k_1 c_B - k_{-1} c_Y$$

and the boundary conditions are equivalent to those already discussed for the EC_{irre} case.

After the normalisation and discretisation of the problem, the equations corresponding to a point i in solution of the linear system to solve are given by

$$\alpha_{i,A} C_{i-1,A}^k + \beta_{i,A}^{\mathrm{dif}} C_{i,A}^k + \gamma_{i,A} C_{i+1,A}^k = \delta_{i,A}^k$$

$$\alpha_{i,B} C_{i-1,B}^k + \left(\beta_{i,B}^{\mathrm{dif}} + K_1 \Delta T\right) C_{i,B}^k + \gamma_{i,B} C_{i+1,B}^k - K_{-1} \Delta T C_{i,Y}^k = \delta_{i,B}^k$$

$$\alpha_{i,Y} C_{i-1,Y}^k + \left(\beta_{i,Y}^{\mathrm{dif}} + K_{-1} \Delta T\right) C_{i,Y}^k + \gamma_{i,Y} C_{i+1,Y}^k - K_1 \Delta T C_{i,B}^k = \delta_{i,Y}^k$$

$$\tag{5.45}$$

---

[5] The simulation of the electrochemical response of the EC_{irre} problem does not require consideration of species Y.

By arranging the unknowns in the way

$$\left[C_{0,A}^k \ C_{0,B}^k \ C_{0,Y}^k \ C_{1,A}^k \ C_{1,B}^k \ C_{1,Y}^k \ldots C_{n-1,A}^k \ C_{n-1,B}^k \ C_{n-1,Y}^k\right]^T \quad (5.46)$$

we find that the corresponding coefficient matrix $\mathbf{A}$ is *heptadiagonal* with the following general structure for a point i in solution:

$$\begin{pmatrix}
\ddots & \ddots & \ddots & \ddots & \ddots & & \ddots & & \ddots \\
\ldots \ 0 & \alpha_{i,A} & 0 & 0 & \beta_{i,A} & 0 & & 0 & \gamma_{i,A} & 0 & 0 \ \ldots \\
\ldots \ 0 & 0 & \alpha_{i,B} & 0 & 0 & \beta_{i,B} & -K_{-1}\Delta T & 0 & \gamma_{i,B} & 0 & \ldots \\
\ldots \ 0 & 0 & 0 & \alpha_{i,Y} & 0 & -K_1\Delta T & \beta_{i,Y} & 0 & 0 & \gamma_{i,Y} \ \ldots \\
& \ddots & \ddots & & \ddots & & \ddots & \ddots & \ddots & \ddots & \ddots
\end{pmatrix} \quad (5.47)$$

For the points situated at the electrode surface ($i = 0$) and bulk solution ($i = n - 1$), the boundary conditions discussed in the previous sections of this chapter apply. The resulting problem can be solved with the algorithm for banded linear equation systems discussed in Section 5.3.

5.4.1. *Testing Simulations: Specific tests*

As illustrated in Figures 5.1 and 5.3, a trivial reference limit to test the simulation of electrochemical systems with homogeneous chemical reactions is that corresponding to very slow chemical kinetics so that the simulated voltammograms must coincide with those of the E mechanism for $K_{1/-1} \to 0$ (see tests discussed in Chapters 3 and 4).

For non-zero chemical kinetics, the chronoamperometric response of the *first-order catalytic mechanism* at a macroelectrode under limiting current conditions is given by [10]:

$$J(T) = -\sqrt{K_1 + K_{-1}} \left(\frac{e^{-\chi}}{\sqrt{\pi\chi}} + \text{erf}(\sqrt{\chi}) \right) \quad (5.48)$$

where

$$\chi = (K_1 + K_{-1}) \, T \quad (5.49)$$

As shown in Figure 5.3, for medium-fast chemical kinetics, the voltammograms of the first-order catalytic mechanism reach steady state conditions where the CV curves are sigmoidal. Such waves are described by the following expression for any eletrode kinetics and $d_B = 1$ [11]

$$J_{ss}(\theta) = -\sqrt{K_1 + K_{-1}} \left[\frac{K_{eq} - e^\theta}{K_{eq}(1 + e^\theta)} \right]$$

$$\times \left(\frac{K_{red} + K_{ox}}{K_{red} + K_{ox} + \sqrt{K_1 + K_{-1}}} \right) \tag{5.50}$$

that in the limit of fully reversible electron transfer, turns into

$$J_{ss}(\theta) = -\sqrt{K_1 + K_{-1}} \frac{K_{eq} - e^\theta}{K_{eq}(1 + e^\theta)} \tag{5.51}$$

Under such steady state conditions, the SWV peak current of the catalytic mechanism at macroelectrodes is given by

$$|\Delta J_{p,ss}| = \sqrt{K_1 + K_{-1}} \tanh\left(\frac{\theta_{SW}}{2}\right) \tag{5.52}$$

For non-catalytic mechanisms, exact solutions under transient conditions are difficult to obtain and their mathematical form is significantly more complex. For a single chemical reaction preceding the electron transfer (*CE mechanism*)

$$Y \underset{k_{-1}}{\overset{k_1}{\rightleftarrows}} A$$

$$A + e^- \rightleftarrows B \tag{5.53}$$

the following expression describes accurately the limiting current chronoamperometry at macroelectrodes when the chemical reaction is not slow (less than 1% error for $(K_1 + K_{-1})T > 10$)

$$|J(T)| = \frac{1 + K_{eq}}{K_{eq}} \frac{\chi}{\sqrt{T}} e^{\chi^2} \mathrm{erfc}(\chi) \tag{5.54}$$

where

$$\chi = \sqrt{(K_1 + K_{-1}) T K_{eq}} \tag{5.55}$$

and $K_{eq} = c_A^*/c_Y^*$. Note that, unlike in the catalytic mechanism, for fast preceding chemical reactions, the current does not become stationary at macroelectrodes and it does not increase continuously with the chemical kinetics but it reaches the following limit value corresponding to fully equilibrium conditions

$$|J(T)| = \frac{1 + K_{eq}}{K_{eq}} \frac{1}{\sqrt{\pi T}} \tag{5.56}$$

In the case of a single chemical reaction following the electron transfer (*EC mechanism*), the limiting current is unaffected by the chemical reaction and it is the position of the voltammetric wave that can be used to test our simulations. This can be performed quantitatively analyzing the normal pulse voltammetry (NPV) that for macroelectrodes, non-slow chemical kinetics, reversible electron transfers ($E_{rev}C$) and only species A being initially present is accurately described by

$$J(T, \theta) = - \left(\frac{1 + K_{eq}}{1 + K_{eq} + e^\theta} \right) \frac{\chi}{\sqrt{T}} e^{\chi^2} \operatorname{erfc}(\chi) \tag{5.57}$$

with

$$\chi = \sqrt{(K_1 + K_{-1})T} \left(\frac{1 + K_{eq} + e^\theta}{K_{eq} e^\theta} \right) \tag{5.58}$$

In the limit of very fast chemical reactions, the value of the half-wave potential of the sigmoidal NPV curve of the $E_{rev}C$ mechanism tends to

$$\theta_{1/2} = \ln\left(1 + K_{eq}\right) \tag{5.59}$$

Under steady state conditions (see Chapter 4), analytical mathematical treatments simplify and easy-to-handle solutions are available for electrodes reactions complicated by coupled chemical reactions. Thus, the following expressions can be employed to test the simulated current-potential response at (hemi)spherical ultramicroelectrodes of the CE mechanism

$$J_{ss}(\theta) = - \frac{\left(1 + K_{eq}\right)\left(1 + \sqrt{K_1 + K_{-1}}\right)}{1 + \left(1 + \sqrt{K_1 + K_{-1}}\right)\left[K_{eq} + (1 + K_{eq}) e^\theta\right]} \tag{5.60}$$

and of the EC mechanism

$$J_{ss}(\theta) = - \frac{\left(1 + K_{eq}\right)\left(1 + \sqrt{K_1 + K_{-1}}\right)}{\left(1 + K_{eq}\right)\left(1 + \sqrt{K_1 + K_{-1}}\right) + \left(1 + K_{eq} + \sqrt{K_1 + K_{-1}}\right) e^\theta} \tag{5.61}$$

5.5. Multiple-Electron Transfer Processes

It is frequent that electroactive molecules exchange more than one electron in successive transfer steps, giving rise to multi-E mechanisms. This situation is commonly found in the electrochemistry of organic and organometallic compounds [3] and biological molecules [12]. In order to show how to tackle this problem, we consider a three-electron process involving four electroactive species, although the same treatment is applicable to any n_r number of consecutive electrochemical processes:

$$
\begin{aligned}
A + e^- &\rightleftarrows B \quad (E^0_{f,1}) \\
B + e^- &\rightleftarrows C \quad (E^0_{f,2}) \\
C + e^- &\rightleftarrows D \quad (E^0_{f,3})
\end{aligned}
\tag{5.62}
$$

where $E^0_{f,n}$ is the formal potential of the nth electron transfer reaction. Note that, depending on the $E^0_{f,n}$ values, homogeneous disproportionation-comproportionation reactions are thermodynamically favoured. For example, for the first two steps the following reaction can be envisaged:

$$
A + C \underset{k_{\text{disp}}}{\overset{k_{\text{comp}}}{\rightleftarrows}} 2B, \quad K_{\text{eq}} = \exp\left[\frac{\left(E^0_{f,1} - E^0_{f,2}\right) F}{\mathcal{R}\mathcal{T}}\right]
\tag{5.63}
$$

Therefore, if $E^0_{f,1} \ll E^0_{f,2}$, the equilibrium constant is very small and so the disproportionation of B into A and C is thermodynamically favourable whereas the opposite applies when $E^0_{f,1} \gg E^0_{f,2}$. The effect of these reactions on the voltammetric response (though not on the concentration profiles) is null in the case of diffusion-only problems when the diffusion coefficients of all the species are the same, the electron transfer processes are fast and there are no other coupled chemical processes [13].[6] Assuming that our experimental system fulfils these conditions, the disproportionation/comproportionation reactions can be ignored such that the differential equation system for each species j is analogous to that of the E mechanism, which for linear diffusion takes the form

$$
\frac{\partial c_j}{\partial t} = D \frac{\partial^2 c_j}{\partial x^2} \quad (j \equiv A, B, C, D)
\tag{5.64}
$$

---

[6] When the experimental system differs significantly from one of these conditions, the effect of disproportionation/comproportionation on the voltammetry can be important and it must be included in the simulation. This case will be discussed in Chapter 6 where second-order chemical kinetic processes are studied.

The surface conditions associated with the kinetics of the different electron transfers can be expressed as

$$D\left(\frac{\partial c_A}{\partial x}\right)_{x=0} = k_{red}^{(1)}c_A(0) - k_{ox}^{(1)}c_B(0)$$

$$D\left(\frac{\partial c_B}{\partial x}\right)_{x=0} = -k_{red}^{(1)}c_A(0) + k_{ox}^{(1)}c_B(0) + k_{red}^{(2)}c_B(0) - k_{ox}^{(2)}c_C(0)$$

$$D\left(\frac{\partial c_C}{\partial x}\right)_{x=0} = -k_{red}^{(2)}c_B(0) + k_{ox}^{(2)}c_C(0) + k_{red}^{(3)}c_C(0) - k_{ox}^{(3)}c_D(0)$$

$$D\left(\frac{\partial c_D}{\partial x}\right)_{x=0} = -k_{red}^{(3)}c_C(0) + k_{ox}^{(3)}c_D(0)$$

(5.65)

such that the conservation of mass principle is fulfilled

$$D\left(\frac{\partial c_A}{\partial x}\right)_{x=0} + D\left(\frac{\partial c_B}{\partial x}\right)_{x=0} + D\left(\frac{\partial c_C}{\partial x}\right)_{x=0} + D\left(\frac{\partial c_D}{\partial x}\right)_{x=0} = 0 \quad (5.66)$$

with $k_{red/ox}^{(n)}$ being the oxidation/reduction rate constants of the nth electron transfer reaction. Considering a two-point approximation for the surface flux, Eqs. (5.65) turn into

$$\frac{C_{1,A}^k - C_{0,A}^k}{h} = K_{red}^{(1)}C_{0,A}^k - K_{ox}^{(1)}C_{0,B}^k$$

$$\frac{C_{1,B}^k - C_{0,B}^k}{h} = -K_{red}^{(1)}C_{0,A}^k + K_{ox}^{(1)}C_{0,B}^k + K_{red}^{(2)}C_{0,B}^k - K_{ox}^{(2)}C_{0,C}^k$$

$$\frac{C_{1,C}^k - C_{0,C}^k}{h} = -K_{red}^{(2)}C_{0,B}^k + K_{ox}^{(2)}C_{0,C}^k + K_{red}^{(3)}C_{0,C}^k - K_{ox}^{(3)}C_{0,D}^k$$

$$\frac{C_{1,D}^k - C_{0,D}^k}{h} = -K_{red}^{(3)}C_{0,C}^k + K_{ox}^{(3)}C_{0,D}^k$$

(5.67)

where $h = X_1 - X_0$.

By arranging the unknowns in the form

$$\left[C_{0,A}^k \ C_{0,B}^k \ C_{0,C}^k \ C_{0,D}^k \ C_{1,A}^k \ C_{1,B}^k \ C_{1,C}^k \ C_{1,D}^k \ \ldots\right]^T \quad (5.68)$$

and assuming that only reactions (5.62) occur in the system, the first four rows of the banded coefficient matrix corresponding to the surface conditions are given by

$$
\begin{pmatrix}
\beta_{0,\mathrm{A}} & -hK_{\mathrm{ox}}^{(1)} & 0 & 0 & \gamma_{0,\mathrm{A}} & 0 & 0 & 0 & 0 & \cdots \\
-hK_{\mathrm{red}}^{(1)} & \beta_{0,\mathrm{B}} & -hK_{\mathrm{ox}}^{(2)} & 0 & 0 & \gamma_{0,\mathrm{B}} & 0 & 0 & 0 & \cdots \\
0 & -hK_{\mathrm{red}}^{(2)} & \beta_{0,\mathrm{C}} & -hK_{\mathrm{ox}}^{(3)} & 0 & 0 & \gamma_{0,\mathrm{C}} & 0 & 0 & \cdots \\
0 & 0 & -hK_{\mathrm{red}}^{(3)} & \beta_{0,\mathrm{D}} & 0 & 0 & 0 & \gamma_{0,\mathrm{D}} & 0 & \cdots \\
\ddots & \ddots & \ddots & \ddots & \ddots & \ddots & \ddots & \ddots & \ddots &
\end{pmatrix}
$$

$$(5.69)$$

where

$$
\begin{aligned}
\beta_{0,\mathrm{A}} &= 1 + hK_{\mathrm{red}}^{(1)}, & \gamma_{0,\mathrm{A}} &= -1 \\
\beta_{0,\mathrm{B}} &= 1 + hK_{\mathrm{ox}}^{(1)} + hK_{\mathrm{red}}^{(2)}, & \gamma_{0,\mathrm{B}} &= -1 \\
\beta_{0,\mathrm{C}} &= 1 + hK_{\mathrm{ox}}^{(2)} + hK_{\mathrm{red}}^{(3)}, & \gamma_{0,\mathrm{C}} &= -1 \\
\beta_{0,\mathrm{D}} &= 1 + hK_{\mathrm{ox}}^{(3)}, & \gamma_{0,\mathrm{D}} &= -1
\end{aligned}
$$

$$(5.70)$$

and $K_{\mathrm{red}}^{(n)} = k_0^{(n)} f^{(n)}(\theta)\,\epsilon/D$ and $K_{\mathrm{ox}}^{(n)} = k_0^{(n)} f^{(n)}(\theta)\,e^{\theta}\epsilon/D$ are the dimensionless reduction and oxidation rate constants of the nth electrode process, respectively (see Chapter 4). Moreover, $\delta_0 = 0$ for all the species. The following rows of the matrix correspond to the discretised Fickian differential equations with the α, β^{dif} and γ coefficients introduced previously.

The above problem can be solved with the algorithm discussed in Section 5.3 and once the concentration profiles are known, the total current is calculated by adding up the contributions of all the n faradaic processes. For the example considered here corresponding to three successive electron transfer steps, and remembering that a negative sign is assigned to the reduction current in this book, this leads to

$$
J = - \left(K_{\mathrm{red}}^{(1)} C_{\mathrm{A}}(0) - K_{\mathrm{ox}}^{(1)} C_{\mathrm{B}}(0) \right) - \left(K_{\mathrm{red}}^{(2)} C_{\mathrm{B}}(0) - K_{\mathrm{ox}}^{(2)} C_{\mathrm{C}}(0) \right)
$$
$$
- \left(K_{\mathrm{red}}^{(3)} c_{\mathrm{C}}(0) - K_{\mathrm{ox}}^{(3)} c_{\mathrm{D}}(0) \right)
$$

$$(5.71)$$

Traditionally, for the sake of accuracy, the calculation of the current is carried out from the surface fluxes of the electroactive species such that the product of large and small numbers is avoided and higher-order approximations for the first derivative at the electrode surface can be employed to increase the accuracy. Thus, attending to Eqs. (5.65), Eq. (5.71) can be rewritten as

$$
J = - \left\{ 3 \left(\frac{\partial C_{\mathrm{A}}}{\partial X} \right)_{X=0} + 2 \left(\frac{\partial C_{\mathrm{B}}}{\partial X} \right)_{X=0} + \left(\frac{\partial C_{\mathrm{C}}}{\partial X} \right)_{X=0} \right\}
$$

$$(5.72)$$

In general, for a series of n_r *consecutive* electron transfers in the form given by (5.62), the current response can be calculated from

$$J = n_r \left(\frac{\partial C_A}{\partial X}\right)_{X=0} + (n_r - 1) \left(\frac{\partial C_B}{\partial X}\right)_{X=0}$$

$$+ (n_r - 2) \left(\frac{\partial C_C}{\partial X}\right)_{X=0} \quad \cdots \quad + \left(\frac{\partial C_N}{\partial X}\right)_{X=0} \tag{5.73}$$

5.5.1. *Testing Simulations: Specific tests*

Depending on the number of electrons transferred and on the thermodynamics and kinetics of such transfers, a wide variety of voltammetric cases can be found in multi-E mechanisms. For the simplest case of two-electron transfers ($n_r = 2$, Figure 5.4), when the second electron transfer is significantly thermodynamically less facile than the first one ($E_{f,2}^0 - E_{f,1}^0 < -200$ mV for electro-reductions at 298.15 K), two well-defined signals can be observed with the features of the CV and SWV curves of reversible monoelectron transfers discussed in Chapters 3 and 4.

As the $E_{f,2}^0 - E_{f,1}^0$ value increases, the two signals gradually overlap. In the particular case $E_{f,2}^0 - E_{f,1}^0 = -35.6$ mV, the peak-to-peak separation in CV and the half-peak width in SWV is the same as for one-electron transfers (Chapters 3 and 4), while the peak height is double in both techniques.

In the second limit situation where the second electron transfer is significantly easier than the first one ($E_{f,2}^0 - E_{f,1}^0 > 200$ mV), a single wave is obtained that in the case of CV is characterised by the following expressions for the dimensionless peak current and peak-to-peak separation at macroelectrodes

$$|J_p| = 2^{3/2} 0.446\sqrt{\sigma} = 1.26\sqrt{\sigma} \tag{5.74}$$

$$|\Delta\theta_p| \approx \frac{2.3}{2} = 1.15 \tag{5.75}$$

and in SWV with small amplitudes the half-peak width is half the value of one-electron transfers

$$W_{1/2}(\theta_{SW} \to 0) = \frac{3.50}{2} = 1.75 \tag{5.76}$$

It is also worth noting that, independently of $E_{f,2}^0 - E_{f,1}^0$, the SWV voltammograms are symmetrical with respect to $\frac{1}{2}\frac{(E_{f,2}^0 + E_{f,1}^0)F}{RT}$.

Fig. 5.4. (a) Cyclic voltammetry and (b) square wave voltammetry of the $E_{rev}E_{rev}$ mechanism with equal diffusion coefficients at a macroelectrode. The grey line corresponds to the response of a simple electron transfer without coupled chemical reactions. (T = 298.15 K)

5.6. Heterogeneous Chemical Processes

In this section we consider the implementation of first-order *heterogeneous* chemical reactions coupled to the electron transfer. In these processes the electroactive species are transformed in a surface-catalysed chemical process that can be characterised through a first-order rate constant k_{het} (m s^{-1}) that is independent of the applied potential. Although a more detailed description of these systems would need to consider the possible adsorption/desorption of the species involved in the heterogeneous chemical transformation,[7] the formalism presented in this section enables a first, simpler description with only one additional unknown parameter: k_{het} [14].

As an example, let us consider the $EC_{irre,het}E$ case that corresponds to the following scheme

$$A + e^- \rightleftarrows B$$

$$B \xrightarrow{k_{het}} C \tag{5.77}$$

$$C + e^- \rightleftarrows D$$

where the coupled chemical process only takes place at the electrode surface and so the variation in the concentrations of the species in solution is given by Fick's second law without additional kinetic terms. On the other hand in the boundary value problem the surface conditions of species B and C must include the kinetics of the chemical process

$$\left. \begin{array}{l} t = 0, \ x \geqslant 0 \\ t > 0, \ x \to \infty \end{array} \right\} \quad c_A = c_A^*, \quad c_B = 0, \quad c_C = 0, \quad c_D = 0 \tag{5.78}$$

$t > 0, \ x = 0:$

$$\begin{aligned} D_A \left(\frac{\partial c_A}{\partial x} \right)_{x=0} &= k_{red}^{(1)} c_A(0) - k_{ox}^{(1)} c_B(0) \\ D_B \left(\frac{\partial c_B}{\partial x} \right)_{x=0} &= -k_{red}^{(1)} c_A(0) + k_{ox}^{(1)} c_B(0) + k_{het} c_B(0) \\ D_C \left(\frac{\partial c_C}{\partial x} \right)_{x=0} &= k_{red}^{(2)} c_C(0) - k_{ox}^{(2)} c_D(0) - k_{het} c_B(0) \\ D_D \left(\frac{\partial c_D}{\partial x} \right)_{x=0} &= -k_{red}^{(2)} c_C(0) + k_{ox}^{(2)} c_D(0) \end{aligned} \tag{5.79}$$

The normalisation of Eq. (5.79) gives rise to

$T > 0, \ X = 0:$

$$\begin{aligned} \left(\frac{\partial C_A}{\partial X} \right)_{X=0} &= K_{red}^{(1)} C_A(0) - K_{ox}^{(1)} C_B(0) \\ \left(\frac{\partial C_B}{\partial X} \right)_{X=0} &= \frac{1}{d_B} \left\{ -K_{red}^{(1)} C_A(0) + \left(K_{ox}^{(1)} + K_{het} \right) C_B(0) \right\} \\ \left(\frac{\partial C_C}{\partial X} \right)_{X=0} &= \frac{1}{d_C} \left\{ K_{red}^{(2)} C_C(0) - K_{ox}^{(2)} C_D(0) - K_{het} C_B(0) \right\} \\ \left(\frac{\partial C_D}{\partial X} \right)_{X=0} &= \frac{1}{d_D} \left\{ -K_{red}^{(2)} C_C(0) + K_{ox}^{(2)} C_D(0) \right\} \end{aligned} \tag{5.80}$$

---

[7] The simulation of adsorption processes will be discussed in Chapter 6.

where the dimensionless rate constant of the heterogeneous chemical reaction is given by

$$K_{\text{het}} = \frac{k_{\text{het}}\epsilon}{D_A} \tag{5.81}$$

According to Eq. (5.80), the first rows of the coefficient matrix of the four-species problem, with the unknowns in the order indicated by (5.68), are given by

$$
\begin{pmatrix}
\beta_{0,A} & -hK_{\text{ox}}^{(1)} & 0 & 0 & \gamma_{0,A} & 0 & 0 & 0 & 0 & \cdots \\
-\frac{1}{d_B}hK_{\text{red}}^{(1)} & \beta_{0,B} & 0 & 0 & 0 & \gamma_{0,B} & 0 & 0 & 0 & \cdots \\
0 & -\frac{1}{d_C}hK_{\text{het}} & \beta_{0,C} & -\frac{1}{d_C}hK_{\text{ox}}^{(2)} & 0 & 0 & \gamma_{0,C} & 0 & 0 & \cdots \\
0 & 0 & -\frac{1}{d_D}hK_{\text{red}}^{(2)} & \beta_{0,D} & 0 & 0 & 0 & \gamma_{0,D} & 0 & \cdots \\
\ddots & & \ddots & & \ddots & & \ddots & & \ddots & \ddots
\end{pmatrix}
\tag{5.82}
$$

where

$$\beta_{0,A} = 1 + hK_{\text{red}}^{(1)}, \qquad\qquad \gamma_{0,A} = -1$$

$$\beta_{0,B} = 1 + \frac{1}{d_B}h\left(K_{\text{ox}}^{(1)} + K_{\text{het}}\right), \quad \gamma_{0,B} = -1$$

$$\beta_{0,C} = 1 + \frac{1}{d_C}hK_{\text{red}}^{(2)}, \qquad\qquad \gamma_{0,C} = -1 \tag{5.83}$$

$$\beta_{0,D} = 1 + \frac{1}{d_D}hK_{\text{ox}}^{(2)}, \qquad\qquad \gamma_{0,D} = -1$$

Regarding the current response, as discussed in Section 5.5, the total current will be given by the sum of the contribution of the different heterogeneous *electrochemical* reactions:

$$
\begin{aligned}
J &= -\left\{ \left(\frac{\partial C_A}{\partial X}\right)_{X=0} + d_C\left(\frac{\partial C_C}{\partial X}\right)_{X=0} + K_{\text{het}}C_B(0) \right\} \\
&= -\left(K_{\text{red}}^{(1)}C_A(0) - K_{\text{ox}}^{(1)}C_B(0) + K_{\text{red}}^{(2)}C_C(0) - K_{\text{ox}}^{(2)}C_D(0) \right) \tag{5.84}
\end{aligned}
$$

References

[1] A. J. Bard and L. R. Faulkner, *Electrochemical Methods: Principles and Applications*. (John Wiley and Sons, New York, 2001).

[2] R. G. Compton and C. E. Banks, *Understanding Voltammetry, 3rd ed.* (World Scientific Press, Singapore, UK, 2018).

[3] O. Hammerich and B. Speiser, *Organic Electrochemistry: Revised and Expanded.* (CRC Press, 2015).

[4] A. Molina, E. Laborda, J. González, and R. G. Compton, Effects of convergent diffusion and charge transfer kinetics on the diffusion layer thickness of spherical micro- and nanoelectrodes, *Phys. Chem. Chem. Phys.* **15**, 7106–7113, (2013).

[5] J. Koutecký and J. Koryta, The general theory of polarographic kinetic currents, *Electrochim. Acta.* **3**, 318–339, (1961).

[6] A. Molina, I. Morales, and M. López-Tenés, Chronoamperometric behaviour of a CE process with fast chemical reactions at spherical electrodes and microelectrodes. Comparison with a catalytic reaction, *Electrochem. Commun.* **8**, 1062–1070, (2006).

[7] D. Britz and J. Strutwolf, *Digital Simulation in Electrochemistry. 4th ed.* (Springer, Berlin, 2016).

[8] F. Martínez-Ortiz, A. Molina, and E. Laborda, Electrochemical digital simulation with highly expanding grid four point discretization: Can Crank-Nicolson uncouple diffusion and homogeneous chemical reactions?, *Electrochim. Acta.* **56**, 5707–5716, (2011).

[9] I. Ruzic and D. Britz, Consistency proof of the sequential algorithm for the digital simulation of systems involving first-order homogeneous kinetics, *Acta Chem. Scand.* **45**, 1087–1089, (1991).

[10] A. Molina, C. Serna, and J. González, General analytical solution for a catalytic mechanism in potential step techniques at hemispherical microelectrodes: Applications to chronoamperometry, cyclic staircase voltammetry and cyclic linear sweep voltammetry, *J. Electroanal. Chem.* **454**, 15–31, (1998).

[11] A. Molina, J. Gómez-Gil, J. Gonzalez, and E. Laborda, Analytical theory for the voltammetry of the non-nernstian catalytic mechanism at macro and microelectrodes: Interplay between the rates of mass transport, electron transfer and catalysis, *Journal of Electroanalytical Chemistry.* 847, 113097, (2019).

[12] P. N. Bartlett, *Bioelectrochemistry: Fundamentals, Experimental Techniques and Applications.* (John Wiley and Sons, New York, 2008).

[13] M. W. Lehmann and D. H. Evans, Effect of comproportionation on voltammograms for two-electron reactions with an irreversible second electron transfer, *Anal. Chem.* **71**, 1947–1950, (1999).

[14] F. Prieto, R. Webster, J. Alden, W. Aixill, G. Waller, R. Compton, and M. Rueda, Electrode processes with coupled chemistry. Heterogeneous or homogeneous chemical reaction? The reduction of nitromethane in basic aqueous solution, *J. Electroanal. Chem.* **437**, 183–189, (1997).

Chapter 6

Second-Order Chemical Kinetic Mechanisms

In the previous chapter coupled chemical reactions that follow first-order chemical kinetics under the experimental conditions were considered. Although the resolution of the problem is more complex since (in general) we deal with coupled equation systems and the spatial and temporal grids may need to be refined when fast chemical processes occur, the mathematical problem nevertheless still consists of *linear* equations.

In this chapter we consider cases where the chemical processes coupled to the electron transfer follow second-order chemical kinetics such that the linearity of the equation system is lost and new mathematical procedures must be employed for the resolution of the corresponding kinetic-diffusion problem. A simple alternative is the linearisation of non-linear terms such that the product of two unknowns (typically concentrations) is expressed as a combination of linear terms. A more general and accurate approach is based on the use of the Newton–Raphson method (or Newton method) for multidimensional root finding [1]. Given that in the electrochemical problems we usually have a rough idea of the range of values of the unknowns (i.e., the concentrations), the efficiency and accuracy of this method is usually very good within the context of electrochemical simulation.

6.1. Second-Order Catalytic Mechanism: The Newton–Raphson Method

To introduce the problematic nature of non-linear equation systems we study the second-order catalytic mechanism where species Y and Z are electroinactive in the range of potentials of the experiments and their concentrations are *not* in large excess compared to the electroactive ones:

$$A + e^- \rightleftharpoons B$$

$$B + Y \underset{k_{-1}}{\overset{k_1}{\rightleftharpoons}} A + Z \tag{6.1}$$

where k_1 and k_{-1} are second-order rate constants with units $M^{-1}s^{-1}$. The corresponding kinetic-diffusion differential equation system under linear diffusion conditions is given by

$$\frac{\partial c_A}{\partial t} = D_A \frac{\partial^2 c_A}{\partial x^2} + k_1 c_B c_Y - k_{-1} c_A c_Z$$

$$\frac{\partial c_B}{\partial t} = D_B \frac{\partial^2 c_B}{\partial x^2} - k_1 c_B c_Y + k_{-1} c_A c_Z$$

$$\frac{\partial c_Y}{\partial t} = D_Y \frac{\partial^2 c_Y}{\partial x^2} - k_1 c_B c_Y + k_{-1} c_A c_Z$$

$$\frac{\partial c_Z}{\partial t} = D_Z \frac{\partial^2 c_Z}{\partial x^2} + k_1 c_B c_Y - k_{-1} c_A c_Z$$

(6.2)

In the initial and bulk conditions we assume that only species A and Y are initially present in solution at the concentrations c_A^* and c_Y^*, respectively:

$$\left. \begin{array}{l} t = 0, \ x \geq 0 \\ t > 0, \ x \to \infty \end{array} \right\} \quad c_A = c_A^*, \quad c_B = 0, \quad c_Y = c_Y^*, \quad c_Z = 0 \qquad (6.3)$$

Otherwise, and assuming that equilibrium conditions apply at the beginning of the experiment, the initial concentrations of the species at the beginning of the simulation are calculated from the corresponding chemical speciation problem.

Regarding the surface conditions, these reflect the electroactivity of the species in the potential range of the study:

$$t > 0, \ x = 0 \quad \begin{cases} D_A \left(\dfrac{\partial c_A}{\partial x} \right)_{x=0} = k_{red} c_A(0) - k_{ox} c_B(0) \\[2mm] D_B \left(\dfrac{\partial c_B}{\partial x} \right)_{x=0} = -D_A \left(\dfrac{\partial c_A}{\partial x} \right)_{x=0} \\[2mm] \left(\dfrac{\partial c_Y}{\partial x} \right)_{x=0} = 0 \\[2mm] \left(\dfrac{\partial c_Z}{\partial x} \right)_{x=0} = 0 \end{cases} \qquad (6.4)$$

The dimensionless form of the problem, by using the parameters and variables introduced in previous chapters ($X = x/\epsilon$, $T = D_A t/\epsilon^2$,

$\theta = \mathrm{F}(E - E_f^0)/\mathcal{R}T$, $K_0 = k_0\epsilon/D_\mathrm{A}$, $d_j = D_j/D_\mathrm{A}$ and $C_j = c_j/c_\mathrm{A}^*$), is given by

$$\frac{\partial C_\mathrm{A}}{\partial T} = \frac{\partial^2 C_\mathrm{A}}{\partial X^2} + K_1^{2\mathrm{nd}}C_\mathrm{B}C_\mathrm{Y} - K_{-1}^{2\mathrm{nd}}C_\mathrm{A}C_\mathrm{Z}$$

$$\frac{\partial C_\mathrm{B}}{\partial T} = d_\mathrm{B}\frac{\partial^2 C_\mathrm{B}}{\partial X^2} - K_1^{2\mathrm{nd}}C_\mathrm{B}C_\mathrm{Y} + K_{-1}^{2\mathrm{nd}}C_\mathrm{A}C_\mathrm{Z}$$

$$\frac{\partial C_\mathrm{Y}}{\partial T} = d_\mathrm{Y}\frac{\partial^2 C_\mathrm{Y}}{\partial X^2} - K_1^{2\mathrm{nd}}C_\mathrm{B}C_\mathrm{Y} + K_{-1}^{2\mathrm{nd}}C_\mathrm{A}C_\mathrm{Z} \tag{6.5}$$

$$\frac{\partial C_\mathrm{Z}}{\partial T} = d_\mathrm{Z}\frac{\partial^2 C_\mathrm{Z}}{\partial X^2} + K_1^{2\mathrm{nd}}C_\mathrm{B}C_\mathrm{Y} - K_{-1}^{2\mathrm{nd}}C_\mathrm{A}C_\mathrm{Z}$$

where the dimensionless form of the second-order rate constants, $K^{2\mathrm{nd}}$, is given by

$$K^{2\mathrm{nd}} = k\frac{c_\mathrm{A}^*\epsilon^2}{D_\mathrm{A}} \tag{6.6}$$

With the central three-point approximation for the spatial derivatives and the backward implicit scheme followed in this book, the differential equations for a point i in solution are

$$\alpha_{i,\mathrm{A}}C_{i-1,\mathrm{A}}^k + \beta_{i,\mathrm{A}}C_{i,\mathrm{A}}^k + \gamma_{i,\mathrm{A}}C_{i+1,\mathrm{A}}^k - K_1^{2\mathrm{nd}}\Delta T C_{i,\mathrm{B}}^k C_{i,\mathrm{Y}}^k + K_{-1}^{2\mathrm{nd}}\Delta T C_{i,\mathrm{A}}^k C_{i,\mathrm{Z}}^k = \delta_{i,\mathrm{A}}^k$$

$$\alpha_{i,\mathrm{B}}C_{i-1,\mathrm{B}}^k + \beta_{i,\mathrm{B}}C_{i,\mathrm{B}}^k + \gamma_{i,\mathrm{B}}C_{i+1,\mathrm{B}}^k + K_1^{2\mathrm{nd}}\Delta T C_{i,\mathrm{B}}^k C_{i,\mathrm{Y}}^k - K_{-1}^{2\mathrm{nd}}\Delta T C_{i,\mathrm{A}}^k C_{i,\mathrm{Z}}^k = \delta_{i,\mathrm{B}}^k$$

$$\alpha_{i,\mathrm{Y}}C_{i-1,\mathrm{Y}}^k + \beta_{i,\mathrm{Y}}C_{i,\mathrm{Y}}^k + \gamma_{i,\mathrm{Y}}C_{i+1,\mathrm{Y}}^k + K_1^{2\mathrm{nd}}\Delta T C_{i,\mathrm{B}}^k C_{i,\mathrm{Y}}^k - K_{-1}^{2\mathrm{nd}}\Delta T C_{i,\mathrm{A}}^k C_{i,\mathrm{Z}}^k = \delta_{i,\mathrm{Y}}^k$$

$$\alpha_{i,\mathrm{Z}}C_{i-1,\mathrm{Z}}^k + \beta_{i,\mathrm{Z}}C_{i,\mathrm{Z}}^k + \gamma_{i,\mathrm{Z}}C_{i+1,\mathrm{Z}}^k - K_1^{2\mathrm{nd}}\Delta T C_{i,\mathrm{B}}^k C_{i,\mathrm{Y}}^k + K_{-1}^{2\mathrm{nd}}\Delta T C_{i,\mathrm{A}}^k C_{i,\mathrm{Z}}^k = \delta_{i,\mathrm{Z}}^k \tag{6.7}$$

where the coefficients α, β, γ and δ^k are those defined in Chapters 3 and 4 for diffusion-only problems.

As anticipated, the equations of the system (6.7) are non-linear since they contain the product of two independent variables. An approximate way of solving the system is the linearisation of the non-linear terms. For example, in the case of (6.7) these correspond to the product of two unknown concentrations at the timestep k that, within the backward implicit method, can be approximated as [2]

$$C_{i,\mathrm{B}}^k C_{i,\mathrm{Y}}^k \approx C_{i,\mathrm{B}}^k C_{i,\mathrm{Y}}^{k-1} + C_{i,\mathrm{B}}^{k-1} C_{i,\mathrm{Y}}^k - C_{i,\mathrm{B}}^{k-1} C_{i,\mathrm{Y}}^{k-1}$$

$$C_{i,\mathrm{A}}^k C_{i,\mathrm{Z}}^k \approx C_{i,\mathrm{A}}^k C_{i,\mathrm{Z}}^{k-1} + C_{i,\mathrm{A}}^{k-1} C_{i,\mathrm{Z}}^k - C_{i,\mathrm{A}}^{k-1} C_{i,\mathrm{Z}}^{k-1} \tag{6.8}$$

where the values of the concentrations of the previous timestep C^{k-1} are known.

A more general and rigorous approach is the determination of the concentrations by employing a multidimensional root-finding algorithm. Indeed, the unknown concentrations can be viewed as the roots of a set of non-linear simultaneous equations to be zeroed, which are given by the discretised differential equations and boundary conditions:

$$f_0 = C_{0,A}^k \left(1 + hK_{\mathrm{red}}\right) - C_{0,B}^k hK_{\mathrm{ox}} - C_{1,A}^k$$
$$f_1 = -C_{0,A}^k \frac{1}{d_B} hK_{\mathrm{red}} + C_{0,B}^k \left(1 + \frac{1}{d_B} hK_{\mathrm{ox}}\right) - C_{1,B}^k$$
$$f_2 = C_{1,Y}^k - C_{0,Y}^k$$
$$f_3 = C_{1,Z}^k - C_{0,Z}^k$$

$$\ldots$$

$$f_{4i} =$$
$$\alpha_{i,A} C_{i-1,A}^k + \beta_{i,A} C_{i,A}^k + \gamma_{i,A} C_{i+1,A}^k - K_1^{\mathrm{2nd}} \Delta T C_{i,B}^k C_{i,Y}^k + K_{-1}^{\mathrm{2nd}} \Delta T C_{i,A}^k C_{i,Z}^k - \delta_{i,A}^k$$

$$f_{4i+1} =$$
$$\alpha_{i,B} C_{i-1,B}^k + \beta_{i,B} C_{i,B}^k + \gamma_{i,B} C_{i+1,B}^k + K_1^{\mathrm{2nd}} \Delta T C_{i,B}^k C_{i,Y}^k - K_{-1}^{\mathrm{2nd}} \Delta T C_{i,A}^k C_{i,Z}^k - \delta_{i,B}^k$$

$$f_{4i+2} =$$
$$\alpha_{i,Y} C_{i-1,Y}^k + \beta_{i,Y} C_{i,Y}^k + \gamma_{i,Y} C_{i+1,Y}^k + K_1^{\mathrm{2nd}} \Delta T C_{i,B}^k C_{i,Y}^k - K_{-1}^{\mathrm{2nd}} \Delta T C_{i,A}^k C_{i,Z}^k - \delta_{i,Y}^k$$

$$f_{4i+3} =$$
$$\alpha_{i,Z} C_{i-1,Z}^k + \beta_{i,Z} C_{i,Z}^k + \gamma_{i,Z} C_{i+1,Z}^k - K_1^{\mathrm{2nd}} \Delta T C_{i,B}^k C_{i,Y}^k + K_{-1}^{\mathrm{2nd}} \Delta T C_{i,A}^k C_{i,Z}^k - \delta_{i,Z}^k$$

$$\ldots$$

$$f_{N-4} = C_{n-1,A}^k - 1$$
$$f_{N-3} = C_{n-1,B}^k$$
$$f_{N-2} = C_{n-1,Y}^k - C_Y^*$$
$$f_{N-1} = C_{n-1,Z}^k$$

$$(6.9)$$

where h is the distance between the first two points of the spatial grid $X_1 - X_0$.

One of the simplest alternatives for multidimensional root finding is the Newton–Raphson method.[1] For the sake of simplicity we introduce it by considering the determination of the root, r, of a single function $f(x)$ such that

$$f(r) = 0 \qquad (6.10)$$

The Newton–Raphson method acts iteratively from an initial guess x_0 that, ideally, is successively improved in the subsequent iterations by correcting the estimate of the previous iteration, x_s, by δx until an imposed accuracy condition is fulfilled; that is, until the new estimate x_{s+1} is close enough to the real root value, r, and then $f(x_{s+1}) \to 0$. From Taylor's expansion the

---

[1] The Newton–Raphson method and other algorithms for multidimensional root-finding can be found in C++ in the section "Multidimensional Root-Finding" of the GNU Scientific Library (GSL): http://www.gnu.org/software/gsl.

value $f(x_s + \delta x)$ can be approximated as

$$f(x_s + \delta x) \approx f(x_s) + f'(x_s)\,\delta x + \frac{f''(x_s)}{2}\delta x^2 + \cdots \tag{6.11}$$

and assuming that the magnitude of the correction term, δx, is small, then the terms of the order δx^2 and higher can be neglected:

$$f(x_s + \delta x) \approx f(x_s) + f'(x_s)\,\delta x \tag{6.12}$$

so that

$$f(x_s + \delta x) = 0 = f(x_s) + f'(x_s)\,\delta x \tag{6.13}$$

and therefore the new guess is given by

$$x_{s+1} = x_s + \delta x = x_s - \frac{f(x_s)}{f'(x_s)} \tag{6.14}$$

The above process is repeated to improve the last guess until the difference between two consecutive estimates, $|\delta x|$, is smaller than a certain *tolerance* level, ε, that is established at the beginning of the process. Another possible criterion is that the value of the function $|f(x_{s+1})|$ is smaller than ε.

This root-finding method is not restricted to one dimension but it can be extended to the resolution of N simultaneous equations of N independent variables. In our case, the $N(= N_p \times N_s)$ variables correspond to the unknown concentrations of N_s species at N_p points of the spatial grid (including the electrode surface and bulk solution boundaries):

$$f_i(x_0, x_1, x_2, \ldots, x_{N-1}) = 0, \quad 0 \leq i \leq (N-1) \tag{6.15}$$

By defining $\mathbf{F}$ as the vector of functions to be zeroed (given by (6.9) in our case), $\mathbf{x}$ as the vector of estimates of the concentrations and $\delta\mathbf{x}$ as the vector with the corrections of the estimates, we can write that

$$f_i(\mathbf{x} + \delta\mathbf{x}) \approx f_i(\mathbf{x}) + \sum_{j=0}^{N-1}\left(\frac{\partial f_i}{\partial x_j}\delta x_j\right) \tag{6.16}$$

which in matrix form is given by

$$\mathbf{F}(\mathbf{x} + \delta\mathbf{x}) = \mathbf{F}(\mathbf{x}) + \mathbf{J}\delta\mathbf{x} \tag{6.17}$$

where $\mathbf{J}$ is the Jacobian $N \times N$ matrix of partial derivatives:

$$\mathbf{J} = \begin{pmatrix} \frac{\partial f_0}{\partial x_0} & \frac{\partial f_0}{\partial x_1} & \frac{\partial f_0}{\partial x_2} & \cdots & \frac{\partial f_0}{\partial x_{N-1}} \\ \frac{\partial f_1}{\partial x_0} & \frac{\partial f_1}{\partial x_1} & \frac{\partial f_1}{\partial x_2} & \cdots & \frac{\partial f_1}{\partial x_{N-1}} \\ \frac{\partial f_2}{\partial x_0} & \frac{\partial f_2}{\partial x_1} & \frac{\partial f_2}{\partial x_2} & \cdots & \frac{\partial f_2}{\partial x_{N-1}} \\ \vdots & \vdots & \vdots & & \vdots \\ \frac{\partial f_{N-1}}{\partial x_0} & \frac{\partial f_{N-1}}{\partial x_1} & \frac{\partial f_{N-1}}{\partial x_2} & \cdots & \frac{\partial f_{N-1}}{\partial x_{N-1}} \end{pmatrix} \tag{6.18}$$

Thus, to obtain the correction term vector $\delta\mathbf{x}$, and then the new estimates, $\mathbf{x} + \delta\mathbf{x}$, the matrix equation to solve is

$$\mathbf{J}\delta\mathbf{x} = -\mathbf{F}(\mathbf{x}) \tag{6.19}$$

Noting the form of the functions in (6.9), and by arranging the $\mathbf{x}$ vector as $x_0 = C_{0,\mathrm{A}}^k$, $x_1 = C_{0,\mathrm{B}}^k$, $x_2 = C_{0,\mathrm{Y}}^k$, $x_3 = C_{0,\mathrm{Z}}^k$, $x_4 = C_{1,\mathrm{A}}^k, \ldots, x_{N-1} = C_{n-1,\mathrm{Z}}^k$, we can infer that the Jacobian matrix of the problem of the second-order catalytic mechanism is nine-diagonal, where the first rows corresponding to the boundary conditions of species A, B, Y and Z are given by

$$\begin{pmatrix} (1 + hK_{\mathrm{red}}) & -hK_{\mathrm{ox}} & 0 & 0 & -1 & 0 & 0 & 0 & 0 & \cdots \\ -\frac{1}{d_{\mathrm{B}}}hK_{\mathrm{red}} & \left(1 + \frac{1}{d_{\mathrm{B}}}hK_{\mathrm{ox}}\right) & 0 & 0 & 0 & -1 & 0 & 0 & 0 & \cdots \\ 0 & 0 & -1 & 0 & 0 & 0 & 1 & 0 & 0 & \cdots \\ 0 & 0 & 0 & -1 & 0 & 0 & 0 & 1 & 0 & \cdots \\ \ddots & & \ddots & & \ddots & \ddots & \ddots & \ddots & \ddots & \ddots \end{pmatrix} \tag{6.20}$$

The following rows correspond to a point i in solution $(1 \leq i \leq n - 2)$ and, for example, for species A have the form given in (6.21) where the coefficients α, β and γ are those defined in the absence of chemical processes in Chapter 4:

$$\ldots \alpha_{i,\mathrm{A}}\ 0\ 0\ 0\ (\beta_{i,\mathrm{A}} + K_{-1}^* x_{4i+3})\ (-K_1^* x_{4i+2})\ (-K_1^* x_{4i+1})\ (K_{-1}^* x_{4i})\ \gamma_{i,\mathrm{A}} \ldots \tag{6.21}$$

where $K_{1/-1}^* = K_{1/-1}^{2\mathrm{nd}}\Delta T$.

Finally, the last four rows corresponding to the point at the bulk solution $i = n - 1$ are given by

$$\begin{pmatrix} \ddots & \ddots & \ddots & \ddots & \ddots & \ddots & \ddots & \ddots & \ddots & \\ \cdots & 0 & 0 & 0 & 0 & 0 & 1 & 0 & 0 & 0 \\ \cdots & 0 & 0 & 0 & 0 & 0 & 0 & 1 & 0 & 0 \\ \cdots & 0 & 0 & 0 & 0 & 0 & 0 & 0 & 1 & 0 \\ \cdots & 0 & 0 & 0 & 0 & 0 & 0 & 0 & 0 & 1 \end{pmatrix} \tag{6.22}$$

The matrix equation (6.19) can be solved by applying the algorithm introduced in Section 5.3 such that a better estimate of the solutions is obtained. The process is repeated until the tolerance criterion is fulfilled. The concentration profiles of the different species will correspond to the solution vector $\mathbf{x} + \delta\mathbf{x}$ and the current response can be calculated from the surface gradient of species A:

$$J = -\left(\frac{\partial C_A}{\partial X}\right)_{X=0} = -[K_{red}C_A(0) - K_{ox}C_B(0)] \qquad (6.23)$$

Figure 6.1 shows the cyclic voltammetry of the second-order catalytic mechanism for different kinetics with $k_1/k_{-1} = 10^6$ and the concentrations of species A and Y being similar. For slow kinetics, the effect of the coupled reaction is analogous to that described for the first-order catalytic mechanism in Chapter 5, the reduction signal of the forward scan increasing, and that of the reverse scan decreasing due to the chemical transformation of species B into A. For fast kinetics, under the conditions of the figure, the second-order catalytic mechanism shows a characteristic split wave in the forward scan [3] so that a prewave at positive overpotentials appears.

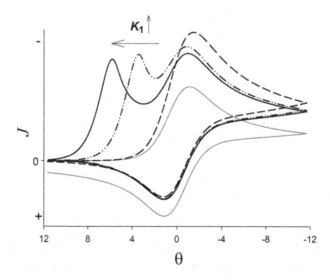

Fig. 6.1. Cyclic voltammetry of the second-order catalytic mechanism (black lines) under linear diffusion conditions and for a fully reversible electrode reaction (grey line). $K_{eq} = 10^6$, $C_Y^* = 1$, $C_B^* = C_Z^* = 0$.

Thus, since Y is not present in a large excess, there is a depletion of its concentration next to the electrode surface and the current of the prewave drops. By continuing the scan, when the applied potential approaches the formal potential of the redox couple A/B, the remaining unreacted species A is electroreduced, giving rise to the second peak.

Note that different limit situations could be taken to test the simulations. Thus, for $K_{1/-1} \to 0$ or $c_Y^* = c_Z^* = 0$, the results must coincide with those expected for the simple E mechanism (Chapters 3 and 4). As the concentrations of species Y and/or Z is increased, the voltammogram must approach to and eventually $(c_Y^*, c_Z^* >> c_A^*, c_B^*)$ converge to that of pseudo-first order kinetics with the corresponding pseudo-first order rate constants (eg., $K_1 = \frac{c_Y^*}{c_A^*} K_1^{2nd}$).

For the application of the Newton–Raphson method detailed above, we must define the functions (as shown in (6.9)), their derivatives (i.e., the Jacobian matrix) and the vector of initial estimates that corresponds to the bulk concentrations of the different species for $k = 0$ and, subsequently, to the concentrations of the previous timestep $k-1$. We also have to provide the tolerance criterion together with a maximum number of iterations such that a convergence error is obtained if the accuracy required is not reached after a reasonable number of cycles. Different criteria can be imposed in a multidimensional problem:

$$\text{All } |f_i(\mathbf{x} + \delta\mathbf{x})| < \varepsilon \tag{6.24}$$

$$\sum_{i=0}^{N-1} |f_i(\mathbf{x} + \delta\mathbf{x})| < \varepsilon \tag{6.25}$$

$$\text{All } |\delta x_i| < \varepsilon \tag{6.26}$$

$$\sum_{i=0}^{N-1} |\delta x_i| < \varepsilon \tag{6.27}$$

The Newton–Raphson method has convergence problems when the initial guess is not good, that is, it is not near the root value. Fortunately, within the context of electrochemical simulation we have a good idea of the range of values for the roots. Thus, the guess at the beginning of the iteration process is the concentration profile from the previous timestep that usually, except for large potential jumps, is not very different from that at the subsequent step. Moreover, for most of the electrochemical systems, the components of the Jacobian matrix can be implemented easily in an analytical form. Therefore, the Newton–Raphson method has proven to work efficiently within the context of the simulation of electrochemical experiments.

6.2. Multiple-Electron Transfers: Adaptive Spatial Grids

In the previous chapter we have introduced the case of multiple-electron transfers (multi-E mechanisms). As discussed, depending on the formal potentials of the different electrochemical steps, comproportionation or disproportionation reactions are thermodynamically favourable. For example, let us consider the case of two consecutive reduction processes (the EE mechanism)

$$
\begin{aligned}
A + e^- &\rightleftarrows B \ (E_{f,1}^0) \\
B + e^- &\rightleftarrows C \ (E_{f,2}^0 \ll E_{f,1}^0)
\end{aligned}
\tag{6.28}
$$

where the formal potential of the second step is much more negative than the first one so that the solution electron-transfer reaction (6.29)

$$
A + C \underset{k_{\text{disp}}}{\overset{k_{\text{comp}}}{\rightleftarrows}} 2\,B
\tag{6.29}
$$

is displaced towards the formation of species B (that is, the comproportionation of A and C is favoured) with the equilibrium constant being given by

$$
K_{\text{eq}} = \exp\left[\frac{F\left(E_{f,1}^0 - E_{f,2}^0\right)}{\mathcal{R}\mathcal{T}}\right]
\tag{6.30}
$$

If the electron transfers are not reversible, the diffusion coefficients of the species are different, there is mass transport by migration or other chemical reactions take place, reaction (6.29) will affect the voltammetry. As an example, Figure 6.2 shows the voltammograms of the EE mechanism where both electron transfers are reversible, diffusion is the only active mechanism, and no side reactions take place *but* the diffusion coefficient of species B is significantly larger than that of species A and C: $D_B = 4\,D_{A/C}$. Under such conditions we can observe that the voltammetry is significantly affected by the kinetics of the comproportionation reaction.

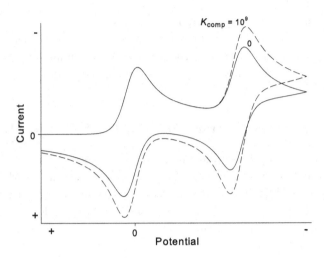

Fig. 6.2. Cyclic voltammetry of the EE mechanism under linear diffusion conditions and for fully reversible electrochemical steps. $E^0_{f,2} - E^0_{f,1} = -0.4$ V, $D_A = D_C = 10^{-5}$ cm^{2}s^{-1}, $D_B = 4 \times 10^{-5}$ cm^{2}s^{-1}.

The corresponding kinetic-diffusion differential equation system to be solved is non-linear:

$$\frac{\partial c_A}{\partial t} = D_A \frac{\partial^2 c_A}{\partial x^2} + k_{disp} c_B^2 - k_{comp} c_A c_C$$

$$\frac{\partial c_B}{\partial t} = D_B \frac{\partial^2 c_B}{\partial x^2} - 2k_{disp} c_B^2 + 2k_{comp} c_A c_C \qquad (6.31)$$

$$\frac{\partial c_C}{\partial t} = D_C \frac{\partial^2 c_C}{\partial x^2} + k_{disp} c_B^2 - k_{comp} c_A c_C$$

with the boundary value problem (assuming that only species A is initially present)

$$\left.\begin{array}{l} t = 0, \ x \geq 0 \\ t > 0, \ x \to \infty \end{array}\right\} \quad c_A = c_A^*, \quad c_B = 0, \quad c_C = 0 \qquad (6.32)$$

$t > 0, \ x = 0$:

$$\begin{cases} D_A \left(\dfrac{\partial c_A}{\partial x}\right)_{x=0} = k_{red}^{(1)} c_A(0) - k_{ox}^{(1)} c_B(0) \\[2mm] D_B \left(\dfrac{\partial c_B}{\partial x}\right)_{x=0} = -k_{red}^{(1)} c_A(0) + k_{ox}^{(1)} c_B(0) + k_{red}^{(2)} c_B(0) - k_{ox}^{(2)} c_C(0) \\[2mm] D_C \left(\dfrac{\partial c_C}{\partial x}\right)_{x=0} = -k_{red}^{(2)} c_B(0) + k_{ox}^{(2)} c_C(0) \end{cases}$$

$$(6.33)$$

After normalisation and discretisation of the above equations, the Newton–Raphson method discussed in the previous section can be employed for the determination of the concentration profiles and the current response that is given by

$$J = -\left\{ 2\left(\frac{\partial C_A}{\partial X}\right)_{X=0} + d_B \left(\frac{\partial C_B}{\partial X}\right)_{X=0} \right\} \qquad (6.34)$$

Within non-linear problems, the present case is particularly interesting because the analysis of the concentration profiles shows that a sharp reaction front develops away from the electrode surface when the comproportionation reaction is very fast (see Figure 6.3). Thus, at negative potentials where species C is formed at the electrode surface the concentration of species A drops sharply due to the fast comproportionation reaction (diffusion limited in the conditions of the figure) and a sharp maximum in the concentration profile of species B is observed. This is time-dependent and moves towards the bulk solution as the scan proceeds.

In this situation, the exponentially expanding grid introduced in Chapter 4 may not place enough points in the region of the front for accurate simulations. Therefore, the spatial grid needs to be refined not only next to the electrode surface but also around the location of the reaction front, X_F in Figure 6.3. This task is challenging considering that, in general, the location of X_F is unknown and time-dependent.

Simple solutions for this problem include the use of very dense grids and static patching schemes. The former is reasonable for one-dimensional problems with modern computers but it can be inefficient in terms of simulation time for multidimensional problems. Patching schemes are more efficient since they enable the use of fewer points. As described in Chapter 4, a high point density can be ensured next to the electrode surface and around a point in solution (X_F in this case) with an expanding-compressing grid in between. Thus, the X_F value can be adjusted for the particular system under study from a preliminary examination of the concentration profiles with a dense spatial grid that enables us to determine the conditions where sharp fronts appear and the region in solution where it moves.

For a more general strategy, the use of *dynamically adaptive grids* has been proposed by several authors [4, 5] as a more suitable and (in some cases) more efficient solution for the problem of moving reaction fronts. The idea underlying it is that the position of the front is automatically identified by the program and the grid is refined accordingly at every timestep. The optimum conditions for the use of these strategies in electrochemical

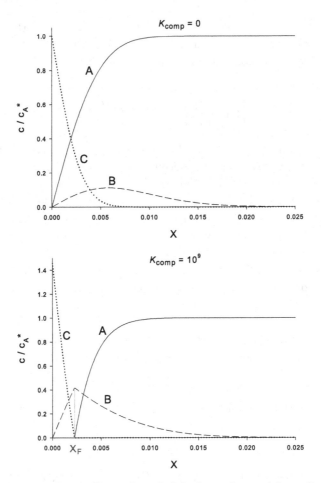

Fig. 6.3.   Concentration profiles at the end of the forward scan of the cyclic voltammo-grams shown in Figure 6.2.

simulation are under study (and debate [6]) due to the difficulty of designing simulation methods that are automatic (without prior knowledge about the concentration profile behaviour), universal (valid for the different reaction mechanisms, kinetics and electrochemical techniques) and still efficient.

Although the implementation and use of adaptive grids are beyond the scope of this book, an overview of the main steps to follow will be given next:[2]

---

[2] Note that the same philosophy can be applied to time discretisation such that more timesteps are employed when the time variation of the concentration profiles is more significant, for example, after a potential jump.

1. *Location of the reaction front*

The first stage requires the programmer to define suitable criteria for the identification of the regions of the spatial grid where more nodes are necessary to obtain accurate values for the concentrations and concentration gradients. With this aim, works in the field have employed the evaluation of concentration gradients [7, 8], comparison of the solutions obtained with different refinement levels [5] or the determination of the local rate of the homogeneous chemical reactions (that is, the magnitude of the kinetic terms of the corresponding kinetic-diffusion equation) [4]. Thus, at each timestep a coarse or exponentially expanding grid is employed at the zero refinement level and the regions where sharp concentration variations lead to unacceptable errors are detected according to the criterion defined.

2. *Regridding*

The grid is refined in those regions that need more spatial resolution, usually employing a uniform mesh for higher accuracy. The *local* problem of each of these regions is then solved with the new local grid and the concentration values at the limit points of the patch (already known from the previous refinement level) as boundary conditions.

The process is repeated until all the points of the concentration profiles fulfil the accuracy criterion. A minimum spatial interval can also be established such that a convergence error is obtained if the accuracy required needs for more refined grids.

3. *Interpolation*

As discussed throughout the previous chapters, for the calculation of the concentration profiles at a given timestep, k, we need the values of the concentrations at the *same* spatial nodes at the previous timestep: $k - 1$. However, because the grids for k and $k-1$ are in general different in adaptive grid techniques, the values C^{k-1} at the nodes of the new grid need to be calculated by suitable interpolation from the old concentration profiles.

As can be inferred from the above discussion, the implementation of adaptive grids is not straightforward, involves additional steps that may make the simulation less efficient and requires the selection of suitable criteria for the location of the moving front. Moreover, the number of experimental systems that present sharp concentration variations away from the electrode surface is quite limited. Therefore, the universality and value of adaptive methods may be called into question, and their use restricted to the development of general simulation packages where the simulation parameters are optimized automatically.

6.3. Adsorption

In this section we consider the possibility that the electroactive species adsorb on the electrode surface (without partial charge transfer) under non-equilibrium conditions. Thus, both species A and B can adsorb on the electrode surface and the electron transfer can take place with both the species in solution ($k_{red/ox,sol}$) and surface-bound ($k_{red/ox,ads}$) as shown in Figure 6.4. The treatment of electrochemical systems with adsorption is significantly more complicated given that we must select a suitable model to describe the adsorption process which will introduce new variables, uncertainties and approximations. Moreover, as will be discussed below, in general these models will lead to non-linear terms in the mathematical problem. For all the above reasons, it is common practice to try to minimise the incidence of adsorption by means of the experimental conditions (mainly the electrode material and solvent). However, in some situations adsorption cannot be avoided (being even intrinsic to the process under study) or it can be desirable as in the modification of electrodes with electroactive monolayers for electroanalysis or electrocatalysis.

When considering the adsorption of the electroactive species A and B, two new independent variables arise that correspond with the amounts of species A, Γ_A (mol m^{-2}), and species B, Γ_B (mol m^{-2}), adsorbed on the surface. Consequently, two new equations are included in the system given by the variation of Γ_A and Γ_B with time as a result of the electron transfer reaction and the adsorption/desorption process (see Figure 6.4):[3]

$$\frac{d\Gamma_A}{dt} = v_{ads,A} - v_{des,A} - k_{red}^{ads}\Gamma_A + k_{ox}^{ads}\Gamma_B$$

$$\frac{d\Gamma_B}{dt} = v_{ads,B} - v_{des,B} + k_{red}^{ads}\Gamma_A - k_{ox}^{ads}\Gamma_A$$

(6.35)

where $k_{red/ox}^{ads}$ (s^{-1}) are the electrochemical rate constants and $v_{ads/des,j}$ the rates of adsorption/desorption of the corresponding species j. Different

---

[3] For the sake of simplicity we assume here that the adsorbed species are not involved in coupled chemical processes. If so, the corresponding kinetic terms must be added in Eq. (6.35).

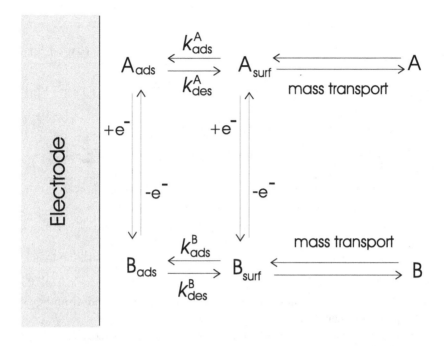

Fig. 6.4.   Reaction scheme including the adsorption of electroactive species.

forms for these rates can be employed. A frequent choice is the Langmuir adsorption isotherm that provides a realistic description of many processes without important additional complications in the numerical simulation. According to this model, the rates of adsorption and desorption with two adsorbing species that compete for the same surface sites are given by

$$v_{\text{ads},j} = k_{\text{ads}}^{j} c_{0,j} \left[\Gamma_{\text{max},j} - (\Gamma_{\text{A}} + \Gamma_{\text{B}}) \right]$$
$$v_{\text{des},j} = k_{\text{des}}^{j} \Gamma_{j}$$
$$(6.36)$$

where k_{ads}^{j} $(\text{mol}^{-1}\text{m}^{3}\text{s}^{-1})$ and k_{des}^{j} (s^{-1}) are the adsorption and desorption rate constants of species j, and $\Gamma_{\text{max},j}$ the maximum surface concentration that we are going to assume to be equal for both species: $\Gamma_{\text{max,A}} = \Gamma_{\text{max,B}} = \Gamma_{\text{max}}$.

The occurrence of adsorption obviously affects the surface condition of species A and B in solution and now this is determined not only by the electron transfer reaction but also by the corresponding adsorption process. Thus, taking into account Eqs. (6.35) and (6.36) the surface fluxes are given by

$$D_A \left(\frac{\partial c_A}{\partial x} \right)_{x=0} = k_{red}^{sol} c_A(0) - k_{ox}^{sol} c_B(0) + k_{ads}^A c_A(0) \left[\Gamma_{max} - (\Gamma_A + \Gamma_B) \right] - k_{des}^A \Gamma_A$$

$$D_B \left(\frac{\partial c_B}{\partial x} \right)_{x=0} = -k_{red}^{sol} c_A(0) + k_{ox}^{sol} c_B(0) + k_{ads}^B c_B(0) \left[\Gamma_{max} - (\Gamma_A + \Gamma_B) \right] - k_{des}^B \Gamma_B$$

$$(6.37)$$

After normalisation, Eqs. (6.35) and (6.37) become

$$\frac{d\xi_A}{dT} = -K_{red}^{ads} \xi_A + K_{ox}^{ads} \xi_B + K_{ads}^A \beta C_{0,A} \left[1 - (\xi_A + \xi_B) \right] - K_{des}^A \beta \xi_A$$

$$\frac{d\xi_B}{dT} = K_{red}^{ads} \xi_A - K_{ox}^{ads} \xi_B + K_{ads}^B \beta C_{0,B} \left[1 - (\xi_A + \xi_B) \right] - K_{des}^B \beta \xi_B$$

$$\left(\frac{\partial C_A}{\partial X} \right)_{X=0} = K_{red}^{sol} C_{0,A} - K_{ox}^{sol} C_{0,B} + K_{ads}^A C_{0,A} \left[1 - (\xi_A + \xi_B) \right] - K_{des}^A \xi_A$$

$$\left(\frac{\partial C_B}{\partial X} \right)_{X=0} = \frac{1}{d_B} \left\{ -K_{red}^{sol} C_{0,A} + K_{ox}^{sol} C_{0,B} + K_{ads}^B C_{0,B} \left[1 - (\xi_A + \xi_B) \right] - K_{des}^B \xi_B \right\}$$

$$(6.38)$$

where T, X, d_j and C_j have been defined in Chapters 2 and 4, and $\xi_j = \Gamma_j/\Gamma_{max}$ is the relative surface coverage.

The dimensionless rate constants for adsorption and desorption are assumed to be potential-independent and given by

$$K_{ads}^j = \frac{k_{ads}^j \Gamma_{max} \epsilon}{D_A}$$

$$K_{des}^j = \frac{k_{des}^j \Gamma_{max} \epsilon}{c_A^* D_A}$$

$$(6.39)$$

and the dimensionless saturation parameter β by

$$\beta = \frac{c_A^* \epsilon}{\Gamma_{max}} \tag{6.40}$$

The dimensionless, potential-dependent electrochemical rate constants of the adsorbed species $K_{red/ox}^{ads}$ are given by

$$K_{red/ox}^{sol} = \frac{k_{red/ox}^{sol} \epsilon}{D_A}$$

$$K_{red/ox}^{ads} = \frac{k_{red/ox}^{ads} \epsilon^2}{D_A}$$

$$(6.41)$$

Generally, the electrochemical rate constants are different for the surface-bound and solution-phase redox couples that, within the Butler–Volmer formalism,[4] means they can have different values of the standard rate constant (k_0^{sol}, k_0^{ads}), transfer coefficient (α^{sol}, α^{ads}) and formal potential ($E_f^{0,sol}$, $E_f^{0,ads}$). Note that the latter are related through the adsorption equilibrium constants, k_{ads}^j/k_{des}^j, as

$$E_f^{0,ads} = E_f^{0,sol} - \frac{\mathcal{R}\mathcal{T}}{F}\ln\left(\frac{k_{ads}^A/k_{des}^A}{k_{ads}^B/k_{des}^B}\right) \tag{6.42}$$

Regarding the initial conditions for the surface coverage, different scenarios can be envisaged depending on whether the electrochemical measurements depart from equilibrium conditions or not. Assuming the former situation and the Langmuir isotherm for the adsorption process the initial conditions of the problem are given by

$$T = 0 \begin{cases} C_A = 1 \\ C_B = C_B^* \\ \xi_A = \dfrac{K_{ads}^A/K_{des}^A}{1 + K_{ads}^A/K_{des}^A + C_B^* K_{ads}^B/K_{des}^B} \\ \xi_B = \dfrac{K_{ads}^B/K_{des}^B}{1 + K_{ads}^A/K_{des}^A + C_B^* K_{ads}^B/K_{des}^B} \end{cases} \tag{6.43}$$

Next, let us consider the discretisation of the new equations of the system by following the two-point approximation for the surface derivatives and the backward implicit scheme:

$$\frac{\xi_A^k - \xi_A^{k-1}}{\Delta T} = -K_{red}^{ads}\xi_A^k + K_{ox}^{ads}\xi_B^k + K_{ads}^A\beta C_{0,A}^k\left[1 - \left(\xi_A^k + \xi_B^k\right)\right] - K_{des}^A\beta\xi_A^k$$

$$\frac{\xi_B^k - \xi_B^{k-1}}{\Delta T} = K_{red}^{ads}\xi_A^k - K_{ox}^{ads}\xi_B^k + K_{ads}^B\beta C_{0,B}^k\left[1 - \left(\xi_A^k + \xi_B^k\right)\right] - K_{des}^B\beta\xi_B^k$$

$$\frac{C_{1,A}^k - C_{0,A}^k}{h} = K_{red}^{sol}C_{0,A}^k - K_{ox}^{sol}C_{0,B}^k + K_{ads}^A C_{0,A}^k\left[1 - \left(\xi_A^k + \xi_B^k\right)\right] - K_{des}^A\xi_A^k$$

$$\frac{C_{1,B}^k - C_{0,B}^k}{h} = \frac{1}{d_B}\left\{-K_{red}^{sol}C_{0,A}^k + K_{ox}^{sol}C_{0,B}^k + K_{ads}^B C_{0,B}^k\left[1 - \left(\xi_A^k + \xi_B^k\right)\right] - K_{des}^B\xi_B^k\right\}$$

$$\tag{6.44}$$

where h is the first spatial interval. The rest of the equations of the system correspond to the mass transport of species A and B in solution and the bulk conditions and so they are equivalent to those of a simple electron transfer process (see Chapters 3 and 4).

---

[4] In the case of surface-bound redox couples the asymmetric Marcus–Hush model needs to be used since the Butler–Volmer model has been found inappropriate for the description of the voltammetry of these systems [9].

As can be seen in (6.44), the third term on the right-hand side of the equations introduces non-linear terms in the problem and its resolution can be carried out with the Newton–Raphson method. Thus, the first rows of the Jacobian matrix, with the variables in the order $x_0 = \xi_A^k$, $x_1 = \xi_B^k$, $x_2 = C_{0,A}^k$, $x_3 = C_{0,B}^k$, $x_4 = C_{1,A}^k$, $x_5 = C_{1,B}^k \ldots$ are given by

$$\mathbf{J} = \begin{pmatrix}
\frac{\partial f_0}{\partial \xi_A^k} & \frac{\partial f_0}{\partial \xi_B^k} & \frac{\partial f_0}{\partial C_{0,A}^k} & 0 & 0 & \cdots \\[6pt]
\frac{\partial f_1}{\partial \xi_A^k} & \frac{\partial f_1}{\partial \xi_B^k} & 0 & \frac{\partial f_1}{\partial C_{0,B}^k} & 0 & \cdots \\[6pt]
\frac{\partial f_2}{\partial \xi_A^k} & \frac{\partial f_2}{\partial \xi_B^k} & \frac{\partial f_2}{\partial C_{0,A}^k} & \frac{\partial f_2}{\partial C_{0,B}^k} & \frac{\partial f_2}{\partial C_{1,A}^k} & 0 & 0 & \cdots \\[6pt]
\frac{\partial f_3}{\partial \xi_A^k} & \frac{\partial f_3}{\partial \xi_B^k} & \frac{\partial f_3}{\partial C_{0,A}^k} & \frac{\partial f_3}{\partial C_{0,B}^k} & 0 & \frac{\partial f_3}{\partial C_{1,B}^k} & 0 & \cdots \\[6pt]
0 & 0 & \frac{\partial f_4}{\partial C_{0,A}^k} & 0 & \frac{\partial f_4}{\partial C_{1,A}^k} & 0 & \frac{\partial f_4}{\partial C_{2,A}^k} & 0 & 0 & \cdots \\[6pt]
0 & 0 & 0 & \frac{\partial f_5}{\partial C_{0,B}^k} & 0 & \frac{\partial f_5}{\partial C_{1,B}^k} & 0 & \frac{\partial f_5}{\partial C_{2,B}^k} & 0 & \cdots \\[6pt]
& & & \ddots & \ddots & \ddots & \ddots & \ddots
\end{pmatrix} \tag{6.45}$$

A simple preliminary elimination step in the equation system is necessary to reduce the Jacobian matrix to pentadiagonal form (that is, to remove the term $\frac{\partial f_3}{\partial \xi_A^k}$):

$$\text{row4} = \text{row4} - \text{row3} \left(\frac{\partial f_3}{\partial \xi_A^k} \middle/ \frac{\partial f_2}{\partial \xi_A^k} \right) \tag{6.46}$$

Subsequently the extended Thomas algorithm discussed in Section 5.3 can be employed to proceed with the iterative Newton–Raphson method and calculate the concentration profiles and $\Gamma_{A/B}$ values. Once these are known, the current response is given by the contributions of *both* electron transfer processes:

$$\begin{aligned}
\frac{I}{FA} &= - \left\{ \left[k_{\text{red}}^{\text{sol}} c_A(0) - k_{\text{ox}}^{\text{sol}} c_B(0) \right] + \left[k_{\text{red}}^{\text{ads}} \Gamma_A(t) - k_{\text{ox}}^{\text{ads}} \Gamma_B(t) \right] \right\} \\[6pt]
&= - \left[D_A \left(\frac{\partial c_A}{\partial x} \right)_{x=0} - \left(\frac{d\Gamma_A}{dt} \right) \right]
\end{aligned} \tag{6.47}$$

which can be rewritten in its dimensionless form as

$$J = - \left\{ \left(\frac{\partial C_A}{\partial X} \right)_{X=0} - \frac{1}{\beta} \left(\frac{d\xi_A}{dT} \right) \right\} \tag{6.48}$$

A complete analysis of the effects of adsorption on the voltammetry is difficult given the number of different situations that can be found, mainly determined by the adsorption strength of the electroactive species, the time scale of the experiment and the electron transfer kinetics. The schematic in Figure 6.5 summarises some of the main features of cyclic voltammograms in the presence of adsorption of the electroactive species. For the sake of simplicity, the electron transfer processes are assumed to be reversible.

When the formal potentials of the adsorbed and non-adsorbed redox couples are similar, a single wave is obtained where the contribution of the electron transfer involving the adsorbed species increases with the scan rate. Thus, there is a transition from diffusional-shaped voltammograms at slow scan rates to adsorptive-shaped at fast scan rates. To understand this behaviour and illustrate the characteristics of adsorptive voltammograms let us consider the response in cyclic voltammetry of a monolayer of species A that undergoes a one-electron, fully reversible electron transfer:

$$J = \mp \frac{\sigma}{\beta} \frac{\exp\left[-\frac{F}{\mathcal{R}\mathcal{T}}\left(E - E_f^{0,\text{ads}}\right)\right]}{\left\{1 + \exp\left[-\frac{F}{\mathcal{R}\mathcal{T}}\left(E - E_f^{0,\text{ads}}\right)\right]\right\}^2} \tag{6.49}$$

where the upper sign refers to the cathodic scan and the lower one to the anodic scan.

As can be seen in Figure 6.5, the signal of an electroactive monolayer is quite different from that of an electroactive species diffusing in solution. Thus, considering a reversible process, the voltammograms are symmetric with respect to $E_f^{0,\text{ads}}$ and the potential axis such that both the reduction and oxidation peaks are situated at $E_f^{0,\text{ads}}$ and show the same peak height:

$$J_{\text{p}} = \mp\frac{1}{4}\frac{\sigma}{\beta} \tag{6.50}$$

The magnitude of the peak scales with the scan rate and not with the square root of the scan rate as for solution-phase redox couples. Consequently, the relative effect of adsorption on voltammetry will increase as the time scale of the experiment decreases.

When one of the species adsorbs much more strongly than the other, the formal potentials will be significantly different according to Eq. (6.42). If the product species B is adsorbed more strongly ($k_{\text{ads}}^{\text{B}}/k_{\text{des}}^{\text{B}} \gg k_{\text{ads}}^{\text{A}}/k_{\text{des}}^{\text{A}}$) the wave of the surface-attached redox couple is shifted towards less negative potentials (the reduction of adsorbed A is energetically easier than that of the solution-phase form) and a prewave is observed. On the other hand, when the adsorption of the reactant species A is much stronger, then $E_f^{0,\text{ads}} \ll E_f^{0,\text{sol}}$ and a postwave is obtained. As discussed above, the pre- and postwaves will be more apparent as the scan rate is increased.

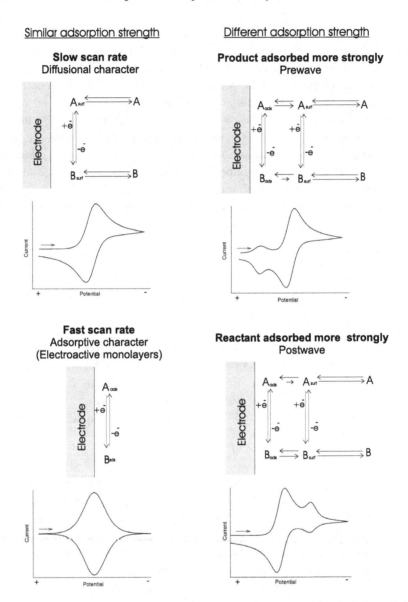

Fig. 6.5.   Different situations in the voltammetry of systems with adsorption of the electroactive species where the electron transfers are assumed to be reversible.

References

[1] W. H. Press, S. A. Teukolsky, W. T. Vetterling, and B. P. Flannery, *Numerical Recipes 3rd edition: The Art of Scientific Computing.* (Cambridge University Press, 2007).

[2] D. Britz and J. Strutwolf, *Digital Simulation in Electrochemistry. 4th ed.* (Springer, Berlin, 2016).

[3] C. P. Andrieux, J. M. Dumas-Bouchiat, and J. M. Savéant, Homogeneous redox catalysis of electrochemical reactions: Part IV. Kinetic controls in the homogeneous process as characterized by stationary and quasi-stationary electrochemical techniques, *J. Electroanal. Chem.* **113**, 1–18, (1980).

[4] C. Amatore, O. Klymenko, and I. Svir, A new strategy for simulation of electrochemical mechanisms involving acute reaction fronts in solution: Principle, *Electrochem. Commun.* **12**, 1170–1173, (2010).

[5] L. K. Bieniasz and C. Bureau, Use of dynamically adaptive grid techniques for the solution of electrochemical kinetic equations: Part 7. Testing of the finite-difference patch-adaptive strategy on example kinetic models with moving reaction fronts, in one-dimensional space geometry, *J. Electroanal. Chem.* **481**, 152–167, (2000).

[6] D. Britz, The true history of adaptive grids in electrochemical simulation, *Electrochim. Acta.* **56**, 4420–4421, (2011).

[7] K. L. Bieniasz, Use of dynamically adaptive grid techniques for the solution of electrochemical kinetic equations: Part 1. Introductory exploration of the finite-difference adaptive moving grid solution of the one-dimensional fast homogeneous reaction-diffusion problem with a reaction layer, *J. Electroanal. Chem.* **360**, 119–138, (1993).

[8] J. G. Blom, J. M. Sanz-Serna, and J. G. Verwer, On simple moving grid methods for one-dimensional evolutionary partial differential equations, *J. Comput. Phys.* **74**, 191–213, (1988).

[9] E. Laborda, M. C. Henstridge, C. Batchelor-McAuley, and R. G. Compton, Asymmetric Marcus-Hush theory for voltammetry, *Chem. Soc. Rev.* **42**, 4894–4905, (2013).

Chapter 7

Electrochemical Simulation in Weakly Supported Media

In many electrochemical measurements a large excess of inert supporting electrolyte is added to the solution containing the species under study. This compresses the electrical double layer (EDL) to a very narrow region around the electrode surface and compensates for the negative (electroreduction) or positive (electrooxidation) charge excess created by electrolysis such that the local solution resistivity is reduced. As a result, the simulation and interpretation of results is greatly simplified since it can be assumed that there is no electric field affecting the concentration profiles of the ionic species that extends from the plane of electron transfer (PET)[1] to bulk solution. The electric potential gradient is restricted to within the EDL so that transport is then exclusively diffusional when macro- and microelectrodes are employed.[2]

The diffusion-only approach yields accurate results for typical experimental conditions where the concentration of the supporting salt is above 0.1 M, which is usually more than 100 times the concentration of the electroactive species [2]. However, under some circumstances the addition of such a large excess of supporting electrolyte is not advisable, as in the case of biological samples, or even impossible, as in non-polar solvents. In other situations the voltammetry with low concentration of supporting electrolyte can provide additional information with respect to diffusion-only voltammetry. For example, as discussed in Chapter 6, conventional voltammetry may be blind to comproportionation reactions taking place in multi-electron transfer processes, whereas they can be characterised in the absence of excess of supporting electrolyte [3]. Finally, understand-

---

[1] The *plane* of electron transfer is, in reality, a *zone* given that the transfer can occur some distance away from the Outer Helmholtz Plane (OHP) via tunnelling [1].

[2] In the case of nanoelectrodes the thickness of the depletion and diffuse layers is typically comparable and then the rigorous simulation of the system must include an appropriate description of the EDL.

ing the electrochemical behaviour of species subject to both diffusional and migrational fields is essential for the progress of nanoelectrochemistry where the diffuse layer penetrates significantly into the depletion one.

From a theoretical point of view, the rigorous treatment of weakly supported media is complex and needs to include several additional aspects with respect to fully supported conditions. First, the negative or positive excess charge generated during electrolysis is not neutralised effectively by the supporting electrolyte ions such that an electric field develops in solution which is more intense in the vicinity of the electrode surface. This will affect the mass transport of the ions in solution as well as the effective driving force for electron transfer at the PET. Moreover, the electrical double layer extends farther into solution and the corresponding electric potential will also affect the concentration profiles of ionic species and the electron transfer kinetics.[3]

In this chapter we focus on the first aspect mentioned above, that is, the electric potential due to the charge generated by electrolysis, which has been found to be the predominant factor that defines the electrochemical response in weakly supported media at macro- and microelectrodes [1, 4]. The latter are usually employed for electrochemical measurements in resistive media given that the dispersion of the charge (i.e., the mass transport of ions) is more efficient and so the results differ less with respect to those predicted for fully supported conditions. The case of macroelectrodes is also interesting in order to establish the experimental conditions under which distorting effects due to the potential drop are not significant. Accordingly, in this chapter we consider the case of (hemi)spherical electrodes that corresponds to a one-dimensional problem and which enables us to move between the limits of linear and steady-state diffusion.

7.1. The Nernst–Planck–Poisson Problem

For the sake of simplicity we will tackle the case of a one-electron transfer (7.1), although the same methodology here described has been successfully applied to the study of a number of systems with different mechanisms, including comproportionation reactions [3] and amalgamation processes [5], and using different electrochemical techniques: chronoamperometry [4], cyclic voltammetry [6] and multipulse voltammetries [7].

---

[3] Note that the value of the formal potential depends on the ionic strength of the solution and so it will also be affected by the concentration of supporting electrolyte.

In the resolution of the problem we consider the electroactive species (A^{z_A} and $B^{z_B = z_A - 1}$) involved in the Faradaic process

$$A^{z_A} + e^- \rightleftharpoons B^{z_A - 1} \tag{7.1}$$

as well as the electroinactive cation, M^{z_M}, and anion, X^{z_X}, present in solution that can proceed from the salt of the electroactive species and/or the added supporting electrolyte. Usually we consider the case MX so that $z_M = -z_X$.

The conversion of species A^{z_A} into $B^{z_A - 1}$ in weakly supported media will give rise not only to a concentration gradient in solution but also to an electric potential gradient. Consequently, a charged species j in solution will be subject to transport by diffusion and migration that is described by the Nernst–Planck equation:

$$j_j = -\left[D_j \left(\frac{\partial c_j}{\partial r} \right) + D_j \frac{z_j F}{RT} c_j \left(\frac{\partial \phi}{\partial r} \right) \right] \tag{7.2}$$

where z_j is the electric charge of species j, j_j the flux and ϕ the electric potential in solution. The first term on the right-hand side corresponds to the diffusional contribution to the flux and the second to the migrational one.

From Eq. (7.2), mass conservation establishes that the variation of the concentration of species j with time and distance to the (hemi)spherical electrode is given by

$$\frac{\partial c_j}{\partial t} = D_j \left(\frac{\partial^2 c_j}{\partial r^2} + \frac{2}{r} \frac{\partial c_j}{\partial r} \right) + D_j \frac{z_j F}{RT} \left(\frac{\partial c_j}{\partial r} \frac{\partial \phi}{\partial r} + c_j \frac{\partial^2 \phi}{\partial r^2} + c_j \frac{2}{r} \frac{\partial \phi}{\partial r} \right) \tag{7.3}$$

Apart from the electroactive and electroinactive species in solution (A, B, M and X), we also need to include the description of the electric potential in the simulation. The Poisson equation relates the potential ϕ with the local electric charge density, ρ (C cm^{-3}):

$$\frac{\partial^2 \phi}{\partial r^2} + \frac{2}{r} \frac{\partial \phi}{\partial r} = -\frac{\rho}{\varepsilon_s \varepsilon_0} \tag{7.4}$$

where ε_s is the relative permittivity of the solvent medium (e.g., *ca.* 80 for water at room temperature), ε_0 the permittivity of free space (8.85×10^{-12} F m^{-1}) and the local charge density is calculated from the concentration and charge of all the ions

$$\rho = F \sum_j z_j c_j \tag{7.5}$$

For the resolution of the above differential equations we need one temporal and two spatial boundary conditions for each unknown, that is, the concentrations of the chemical species and the electric potential. As discussed in previous chapters, the temporal condition corresponds to the beginning of the experiment ($t = 0$) whereas the spatial ones correspond to the electrode surface ($r = r_e$) and the bulk solution ($r \to \infty$). Thus, for the chemical species the boundary value problem reflects whether they are initially present in solution and whether they are electroactive or electroinert in the potential range of the study:

$$
\begin{array}{c}
t = 0, \ r \geq r_e \\
t > 0, \ r \to \infty
\end{array}
\left\{
\begin{array}{c}
c_A = c_A^*, \ c_B = 0 \\
c_M = c_{sup}^*, \ c_X = c_{sup}^* + z_A c_A^* \quad \text{if } z_A > 0 \\
\text{or} \\
c_M = c_{sup}^* - z_A c_A^*, \ c_X = c_{sup}^* \quad \text{if } z_A < 0
\end{array}
\right.
\tag{7.6}
$$

$t > 0, \ r = r_e$:

$$
\left\{
\begin{array}{l}
D_A \left(\dfrac{\partial c_A}{\partial r} \right)_{r=r_e} + \dfrac{z_A F}{RT} D_A c_A \left(\dfrac{\partial \phi}{\partial r} \right)_{r=r_e} = k_{red} c_A(r_e) - k_{ox} c_B(r_e) \\[2mm]
D_B \left(\dfrac{\partial c_B}{\partial r} \right)_{r=r_e} + \dfrac{(z_A - 1) F}{RT} D_B c_B \left(\dfrac{\partial \phi}{\partial r} \right)_{r=r_e} = -D_A \left(\dfrac{\partial c_A}{\partial r} \right)_{r=r_e} - \dfrac{z_A F}{RT} D_A c_A \left(\dfrac{\partial \phi}{\partial r} \right)_{r=r_e} \\[2mm]
D_M \left(\dfrac{\partial c_M}{\partial r} \right)_{r=r_e} + \dfrac{z_M F}{RT} D_M c_M \left(\dfrac{\partial \phi}{\partial r} \right)_{r=r_e} = 0 \\[2mm]
D_X \left(\dfrac{\partial c_X}{\partial r} \right)_{r=r_e} + \dfrac{z_X F}{RT} D_X c_X \left(\dfrac{\partial \phi}{\partial r} \right)_{r=r_e} = 0
\end{array}
\right.
\tag{7.7}
$$

where c_{sup} is the added concentration of supporting electrolyte MX and we have assumed that the counterion of species A is M if $z_A < 0$ or X if $z_A > 0$. Finally, the current can be calculated from the surface flux of the reactant species A:

$$
\frac{I}{FA} = -\left\{ D_A \left(\frac{\partial c_A}{\partial r} \right)_{r=r_e} + \frac{z_A F}{RT} D_A c_A \left(\frac{\partial \phi}{\partial r} \right)_{r=r_e} \right\}
\tag{7.8}
$$

It is important to highlight that the reduction/oxidation rate constants in (7.7), $k_{red/ox}$, are defined by the *real* driving force experienced by the electroactive species, which differs from the applied potential, E, under weakly supported conditions as can be seen in Figure 7.1. Because of the charge excess that builds up adjacent to the working electrode as a result of the electrode reaction, the potential at the tunnelling distance is different from that at the bulk solution. Comparing both graphs in

Fig. 7.1.   Cyclic voltammetry (left) and time variation of the potential difference between the working electrode and the electron transfer place (right) for an electron transfer under fully supported (solid line) and weakly supported (dashed line) conditions.

Figure 7.1 we can infer that, during the electroreduction of A, the negative charge accumulated near the electrode gives rise to a more negative potential at the PET than in bulk solution and so a smaller driving force for reduction: $(\phi_{WE} - \phi_{PET}) > (\phi_{WE} - \phi_{bulk})(= E)$. The opposite situation is found when species B is oxidised to A such that, at some point, a positive charge excess appears adjacent to the working electrode and $(\phi_{WE} - \phi_{PET}) < (\phi_{WE} - \phi_{bulk})$. Consequently there is a loss in the driving force for electron transfer such that the overpotential that the electroactive species "feels" (dashed line in Figure 7.1) is, in general, smaller in absolute value than the applied one (solid line in Figure 7.1). Within the Butler–Volmer formalism, this situation can be expressed as

$$k_{red} = k_0 \exp\left[-\frac{\alpha F}{RT} \left(E - E_f^0 - \Delta\phi\right) \right]$$

$$k_{ox} = k_0 \exp\left[\frac{(1-\alpha)F}{RT} \left(E - E_f^0 - \Delta\phi\right) \right]$$

(7.9)

where $\Delta\phi = \phi_{PET} - \phi_{bulk}$ is the difference between the electric potential at the point in solution immediately adjacent to the electrode surface (at the edge of the double layer) and the bulk solution. Therefore, $\Delta\phi$ quantifies the loss of driving force.

Regarding the boundary conditions for the electric potential, initially uniform concentration profiles are assumed and a null value is assigned to the corresponding uniform potential profile: $\phi(r, t = 0) = 0$. During the experiments, electroneutrality is maintained in bulk solution. Consequently, the potential drop tends towards zero as r tends to infinity but this occurs *beyond* the depletion layer which thus requires the extension of the diffusion-only simulation space (and then the simulation runtime) or the transformation of the space coordinate [4]:

$$y = 1 - \frac{r_e}{r} \qquad (7.10)$$

such that $y = 0$ corresponds to the electrode surface and $y = 1$ to an infinite distance where $\phi(y = 1) = 0$. Another alternative that will be employed here is the partial resolution of the Poisson equation for $r_{max} < r < \infty$ where the solution is electroneutral. As discussed in previous chapters, r_{max} is usually placed in the simulation space at $r_e + 6\sqrt{D_{max} t_{max}}$. Accordingly, Eq. (7.4) shows that

$$\frac{\partial^2 \phi}{\partial r^2} + \frac{2}{r}\frac{\partial \phi}{\partial r} = 0 \qquad (7.11)$$

and, considering that $\phi = 0$ for $r \to \infty$, it leads to the following outer-boundary condition for the electric potential:

$$r_{max}\left(\frac{\partial \phi}{\partial r}\right)_{r=r_{max}} + \phi\left(r_{max}\right) = 0 \qquad (7.12)$$

With respect to the surface condition for ϕ, the zero-field approximation (Eq. (7.13)) introduced by Streeter and Compton [4] has proven to provide accurate results when the size of the electrode is not at the nanoscale ($R_e \geq 10^3$ [8], see Eq. (7.16)):

$$\left(\frac{\partial \phi}{\partial r}\right)_{r=r_e} = 0 \qquad (7.13)$$

This condition implies that the electrical double layer has a negligible thickness with respect to the depletion layer, and that the charge within it balances that of the electrode (see Figure 7.2) such that the electric field is null at the edge of the EDL corresponding to the electron transfer distance: $r = r_{PET}$ (corresponding to r_e in the simulation space within the present treatment). Therefore, a detailed description of the double layer is not necessary (unless the dimensions of the electrode are nanometric), which greatly simplifies the simulation and makes it more efficient since less dense spatial and temporal grids are employed.

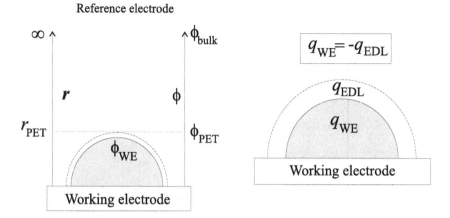

Fig. 7.2. Schematic of the system (left) and the zero-field approximation (right).

Next, let us consider the normalisation of the Nernst–Planck–Poisson system of equations according to the definitions already introduced in previous chapters ($R = r/r_e$, $T = D_A t/r_e^2$, $\theta = F(E - E_f^0)/\mathcal{RT}$, $K_0 = k_0 r_e/D_A$, $d_j = D_j/D_A$ and $C_j = c_j/c_A^*$ with $j = A, B, M, X$):

$$\frac{\partial C_j}{\partial T} = d_j \left(\frac{\partial^2 C_j}{\partial R^2} + \frac{2}{R} \frac{\partial C_j}{\partial R} \right) + d_j z_j \left(\frac{\partial C_j}{\partial R} \frac{\partial \Phi}{\partial R} + C_j \frac{\partial^2 \Phi}{\partial R^2} + C_j \frac{2}{R} \frac{\partial \Phi}{\partial R} \right)$$

$$\frac{\partial^2 \Phi}{\partial R^2} + \frac{2}{R} \frac{\partial \Phi}{\partial R} = -R_e^2 \sum_j z_j C_j$$

$$(7.14)$$

where the dimensionless potential is defined as

$$\Phi = \phi \frac{F}{\mathcal{RT}} \tag{7.15}$$

and

$$R_e = r_e \sqrt{\frac{F^2 c_A^*}{\mathcal{RT} \varepsilon_s \varepsilon_0}} \tag{7.16}$$

which represents the relative scale of the electrode compared to the Debye length.

The dimensionless form of the boundary conditions, taking into account the zero-field approximation ($(\partial \phi / \partial r)_{r=r_e} = 0$), is given by

$T = 0, \ R \geq 1$

$T > 0, \ R = R_{max} = 1 + 6\sqrt{d_{max}T_{max}}$

$$\begin{cases} C_A = 1 \\ C_B = 0 \\ \Phi(T = 0) = 0 \\ \Phi(R = R_{max}) + R_{max}\left(\dfrac{\partial\Phi}{\partial R}\right)_{R=R_{max}} = 0 \\ C_M = C^*_{sup}, \ C_X = C^*_{sup} + z_A \quad \text{if } z_A > 0 \\ \qquad\qquad \text{or} \\ C_M = C^*_{sup} - z_A, \ C_X = C^*_{sup} \quad \text{if } z_A < 0 \end{cases} \qquad (7.17)$$

$T > 0, \ R = 1$:

$$\begin{cases} \left(\dfrac{\partial C_A}{\partial R}\right)_{R=1} = K_{red}C_A(1) - K_{ox}C_B(1) \\ d_B\left(\dfrac{\partial C_B}{\partial R}\right)_{R=1} = -\left(\dfrac{\partial C_A}{\partial R}\right)_{R=1} \\ \left(\dfrac{\partial\Phi}{\partial R}\right)_{R=1} = 0 \\ \left(\dfrac{\partial C_M}{\partial R}\right)_{R=1} = 0 \\ \left(\dfrac{\partial C_X}{\partial R}\right)_{R=1} = 0 \end{cases} \qquad (7.18)$$

where $C^*_{sup} = c^*_{sup}/c^*_A$ is the support ratio of the electrolytic solution. And the dimensionless current is given by

$$J = \frac{I\,r_e}{FAD_Ac^*_A} = -\left(\frac{\partial C_A}{\partial R}\right)_{R=1} \qquad (7.19)$$

The discretisation of the derivatives of the new unknown (the electric potential, ϕ) can be treated in the same way as detailed in previous chapters for the chemical species concentrations such that

$$\begin{aligned} \frac{\partial\Phi}{\partial R} &\approx \frac{\Phi^k_{i+1} - \Phi^k_{i-1}}{\Delta R_+ + \Delta R_-} \\[2mm] \frac{\partial^2\Phi}{\partial R^2} &\approx \frac{\frac{\Phi^k_{i+1}-\Phi^k_i}{\Delta R_+} - \frac{\Phi^k_i-\Phi^k_{i-1}}{\Delta R_-}}{\frac{1}{2}[\Delta R_+ + \Delta R_-]} \end{aligned} \qquad (7.20)$$

It is also interesting to consider the discretisation of the surface condition for the electroactive species that is affected by the potential gradient in solution:

$$\frac{C_{1,A}^k - C_{0,A}^k}{h} = K_0 \exp\left[-\alpha\left(\theta - \Phi_0^k\right)\right] C_{0,A}^k - K_0 \exp\left[(1-\alpha)\left(\theta - \Phi_0^k\right)\right] C_{0,B}^k$$

$$(7.21)$$

where $\theta = \mathrm{F}(E - E_f^0)/\mathcal{R}\mathcal{T}$ and Φ_0^k is the value of the dimensionless electric potential at the electron transfer distance, $i = 0$.

As can be concluded from Eqs. (7.14) and (7.21), the diffusion-migration problem is non-linear. The Newton–Raphson method has been applied successfully to the resolution of the Nernst–Planck–Poisson equation system although the convergence is slower than for the kinetic-diffusion problems studied in Chapter 6. The unknown vector $\mathbf{x}$ corresponds to

$$\mathbf{x} = \left[C_{0,A}^k, C_{0,B}^k, \Phi_0^k, C_{0,M}^k, C_{0,X}^k, \ldots, C_{n-1,A}^k, C_{n-1,B}^k, \Phi_{n-1}^k, C_{n-1,M}^k, C_{n-1,X}^k\right]^T$$

$$(7.22)$$

and the function vector contains $f_0 = f_{0,A}$, $f_1 = f_{0,B}$, $f_2 = f_{0,\Phi}$, $f_3 = f_{0,M}$, $f_4 = f_{0,X}$, $f_5 = f_{1,A}, \ldots, f_{5n-1} = f_{(n-1),X}$ where the functions $f_{0,j}$ correspond to the surface conditions of species j (Eq. (7.18)), $f_{1,j \leq i, j < (n-1),j}$ arise from the discretised form of the Nernst–Planck–Poisson equation system (7.14) and $f_{(n-1),j}$ from the bulk conditions (Eq. (7.17)). The Jacobian matrix resulting from this set of equations and ordering of unknowns and functions is banded and it can be solved with the extended Thomas method described in Chapter 5.

7.2. Weakly Supported Cyclic Voltammetry and Chronoamperometry

Following the numerical procedure discussed in the above section, the electrochemical response of a one-electron reduction process in resistive media will be simulated and studied at macro- and microelectrodes.[4]

In order to understand the deviations with respect to conventional experiments caused by low levels of supporting electrolyte we have to bear in mind that these are the result of the effects of the electric field in solution

---

[4] In the case of nanoelectrodes, it is necessary to include an appropriate description of the double layer in the simulations. This is not straightforward due to the need to include a suitable theoretical model (and therefore new parameters) and a much denser spatial grid [1, 9].

and the ability of the system to disperse it. The former includes the influence of the potential gradient on the mass transport of charged species as well as on the electron transfer kinetics (Eqs. (7.9)). With respect to the neutralisation of the charge excess, this will be more efficient as the concentration and charge of the ions in solution increase and their mass transport is faster, which depends on the species mobilities, the working electrode size and the time scale of the experiment.

Figure 7.3 shows the chronoamperograms and the profile of the potential drop at a spherical microelectrode under weakly supported conditions ($C^*_{sup} = 0.1$) for different z_A values[5] and an applied potential of -0.5 V. This corresponds to mass-transport-limited current conditions in large excess of supporting electrolyte.

As can be observed, the largest distortions at $C^*_{sup} = 0.1$ with respect to the current-time response in excess of supporting electrolyte (grey curve) are found for $z_A = 0$ and $+1$. This can be understood considering the charge and concentration of ions in solution. Thus, when the electroactive species is highly charged, this species and its counterion contribute to the compensation of the electric charge acting themselves as supporting electrolyte (*self-support* conditions). Obviously, the self-support capability of the system decreases with $|z_A|$ so that in the case of neutral species only the added supporting electrolyte (apart from other possible impurities present in the system) is responsible for the reduction of the solution resistivity.

Different regions can be distinguished in the chronoamperograms for $z_A = 0$ and $+1$ in Figure 7.3. At short times the response is mainly controlled by the so-called *ohmic drop effects*, that is, by the loss in the driving force of electron transfer. As time proceeds, the charge excess is gradually dispersed and the potential drop decreases. At some time the potential difference between the working electrode and the PET is large enough for immediate reduction of A at the electrode surface: $c_A(r_e) = 0$. Thus, the chronoamperograms transit to a mass-transport-controlled regime where the variation of current with time is analogous to that in fully supported conditions but its magnitude differs due to the contribution of electromigration to the flux of species A towards the electrode surface. Given that we are considering an electroreduction process where a negative charge is injected in solution, cations will be attracted towards

---

[5] Note that within the context of weakly supported media the results obtained for the electroreduction of A^{+z} are equivalent (except for the sign of the current) to the electrooxidation of A^{-z}.

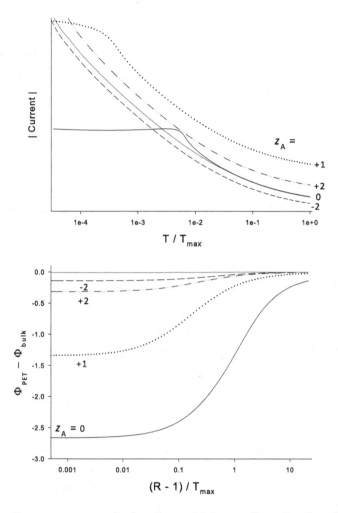

Fig. 7.3. Chronoamperograms (top) and potential drop profile at $T = T_{max}$ (bottom), for the one-electron reduction of species A^{z_A} at a 10 μm-radius spherical electrode when applying an overpotential of -0.5 V ($\theta = -19.5$ at 298.15 K) under fully supported (grey line) and weakly supported ($C_{sup}^* = 0.1$, black lines) conditions. $k_0 = 1$ cm s^{-1}, $D = 10^{-5}$ cm^{2}s^{-1}.

the electrode surface whereas anions will be repelled. Accordingly, in the mass-transport-controlled regime in Figure 7.3 (longer times), the current is larger than the diffusion-only one for $z_A > 0$ and smaller for $z_A < 0$. For neutral species the current tends to that under fully supported conditions since the electroactive species does not migrate.

All the above observations are consistent with the potential drop observed in the bottom graph of Figure 7.3. This shows that in all cases there is a potential drop in solution that reduces the potential difference experienced by species A and attracts or repels it from the electrode surface. This drop is particularly significant for neutral and monocharged species.

The effects of the concentration of supporting electrolyte on the cyclic voltammograms at macroelectrodes (linear diffusion) and under near steady-state conditions are shown in Figure 7.4 for the one-electron reduction of species A with $z_A = +2$, 0 and -1. As occurred with the chronoamperograms at short times, the voltammetry at macroelectrodes is mainly affected by ohmic drop effects as the support ratio, C^*_{sup}, is decreased. Independently of the z_A value, the peak currents of both the cathodic and anodic peaks decrease with C^*_{sup} and they shift towards greater overpotentials. These distortions are the consequence of the discrepancy between the applied potential and the real potential difference at the electron transfer plane (see Figure 7.1). The real driving force is smaller and then mass-transport-controlled conditions are reached at larger potentials. As a result, the peak-to-peak separation increases significantly. Although qualitatively the effects are similar for all z_A values, we can see that these are more apparent in the case of neutral species $z_A = 0$ due to the reasons previously discussed.

When microelectrodes are employed (right-hand graphs in Figure 7.4), the shape of the voltammograms is not significantly distorted with respect to fully supported conditions and these show the characteristic sigmoidal wave even for $C^*_{sup} \to 0$. The predominant effect on the voltammetry corresponds to the increase or decrease in the steady-state limiting current according to the contribution of the transport by migration to the flux of electroactive species A^{z_A}. The values of the change in the steady-state limiting current under self-support conditions and the assumption of electroneutrality [10] were reported by Cooper *et al.* [11] and are given in Table 7.1.

Between the extremes of linear diffusion at macroelectrodes and steady-state voltammetry, intermediate situations can be found where the overall effect on the electrochemical response is the balance of ohmic drop and migration effects. As shown above, in cyclic voltammetry the former increases the peak-to-peak separation and decreases the magnitude of the peaks whereas electromigration can result in the increase or decrease in the current depending on the charge of the electroactive species. Equivalently, in the case of differential pulse voltammetries such as square wave

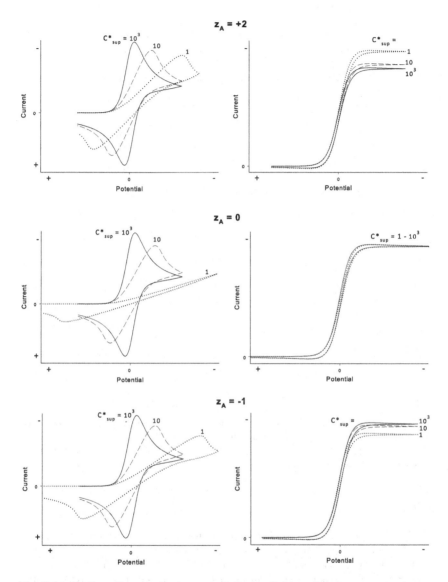

Fig. 7.4. Cyclic voltammetry at macroelectrodes (left) and under near steady-state conditions (right) for the reversible one-electron reduction of species A^{z_A} at different concentrations of supporting electrolyte (C^*_{sup} values indicated on the graphs).

voltammetry (Figure 7.5), ohmic drop effects lead to smaller and broader peaks (Figure 7.5(a)) and transport by migration affects the peak current according to the z_A–value (see Figure 7.5(b)).

Table 7.1. Ratio between the steady-state limiting current under self-support and fully supported conditions $(I_{\lim}/I_{\lim,\mathrm{dif}})$ for the one-electron reduction of species A with $D_A = D_B = D_M = D_X$, $Z_M = +1$, $Z_X = -1$.

z_A	$I_{\lim}/I_{\lim,\mathrm{dif}}$	z_A	$I_{\lim}/I_{\lim,\mathrm{dif}}$
+4	1.128	−1	0.849
+3	1.173	−2	0.880
+2	1.274	−3	0.902
+1	2.000	−4	0.918
0	1.000		

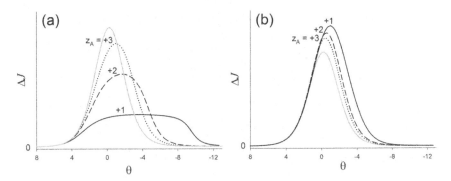

Fig. 7.5. Square wave voltammetry at (a) macroelectrodes and (b) near steady-state conditions of the reversible one-electron reduction of species A^{z_A} for different z_A–values under weakly supported conditions $(C^*_{\mathrm{sup}} = 0.1)$. The grey line curves correspond to fully-supported conditions.

References

[1] E. J. F. Dickinson and R. G. Compton, Influence of the diffuse double layer on steady-state voltammetry, *J. Electroanal. Chem.* **661**, 198–212, (2011).

[2] E. J. F. Dickinson and R. G. Compton, How much supporting electrolyte is required to make a cyclic voltammetry experiment quantitatively "diffusional"? A theoretical and experimental investigation, *J. Phys. Chem. C.* **113**, 11157–11171, (2009).

[3] S. R. Belding, J. G. Limon-Petersen, E. E. J. F. Dickinson, and R. G. Compton, Cyclic voltammetry in the absence of excess supporting electrolyte offers extra kinetic and mechanistic insights: Comproportionation of anthraquinone and the anthraquinone dianion in acetonitrile, *Angew. Chem. Int. Ed.* **49**, 9242–9245, (2010).

[4] I. Streeter and R. G. Compton, Numerical simulation of potential step chronoamperometry at low concentrations of supporting electrolyte, *J. Phys. Chem. C.* **112**, 13716–13728, (2008).

[5] J. G. Limon-Petersen, I. Streeter, N. V. Rees, and R. G. Compton, Voltammetry in weakly supported media: The stripping of thallium from a hemispherical amalgam drop. Theory and experiment, *J. Phys. Chem. C.* **112**, 17175–17182, (2008).

[6] J. G. Limon-Petersen, E. J. F. Dickinson, S. R. Belding, N. V. Rees, and R. G. Compton, Cyclic voltammetry in weakly supported media: The reduction of the cobaltocenium cation in acetonitrile. Comparison between theory and experiment, *J. Electroanal. Chem.* **650**, 135–142, (2010).

[7] Y. Wang, E. O. Barnes, E. Laborda, A. Molina, and R. G. Compton, Differential pulse techniques in weakly supported media. Changes in the kinetics and thermodynamics of electrode processes resulting from the supporting electrolyte concentration, *J. Electroanal. Chem.* **673**, 13–23, (2012).

[8] E. J. F. Dickinson and R. G. Compton, The zero-field approximation for weakly supported voltammetry: A critical evaluation, *Chem. Phys. Lett.* **497**, 178–183, (2010).

[9] Y. Sun, Y. Liu, Z. Liang, L. Xiong, A. Wang, and S. Chen, On the applicability of conventional voltammetric theory to nanoscale electrochemical interfaces, *J. Phys. Chem. C.* **113**(22), 9878–9883, (2009).

[10] E. J. Dickinson, J. G. Limon-Petersen, and R. G. Compton, The electroneutrality approximation in electrochemistry, *Journal of Solid State Electrochemistry.* **15**(7-8), 1335–1345, (2011).

[11] J. B. Cooper, A. M. Bond, and K. B. Oldham, Microelectrode studies without supporting electrolyte: Model and experimental comparison for singly and multiply charged ions, *J. Electroanal. Chem.* **331**, 877–895, (1992).

Chapter 8

Hydrodynamic Voltammetry

Electrochemical systems where the mass transport of chemical species is due to diffusion and electromigration were studied in previous chapters. In the present chapter, we are going to consider the occurrence of the third mechanism of mass transfer in solution: convection. Although the modelling of natural convection has experienced some progress in recent years [1], this is usually avoided in electrochemical measurements. On the other hand, convection applied by an external source (*forced convection*) is employed in valuable and popular electrochemical methods in order to enhance the mass transport of species towards the electrode surface. Some of these *hydrodynamic methods* are based on electrodes that move with respect to the electrolytic solution, as rotating electrodes [2], whereas in other hydrodynamic systems the electrolytic solution flows over a static electrode, as in wall-jet [3] and channel electrodes [4].

Although hydrodynamic systems are more difficult to fabricate and simulate than diffusion-only ones, they offer important advantages in electroanalysis and mechanistic studies. The rate of mass transfer can be varied over a large range and tuned in order to establish the optimum conditions for the study of homogeneous and heterogeneous kinetics. Thus, faster kinetics can be studied by increasing the mass transport through the convective flux. The more efficient mass transport also enables us to attain greater sensitivity and steady-state responses that are not affected by the double-layer charging and are easier to simulate. Moreover, some of these methods are ideal for the monitoring of flowing samples with electrochemical measurements.

When both diffusion and convection act in solution, the flux of a species j is given by

$$\mathbf{j}_j = -D_j \boldsymbol{\nabla} c_j + c_j \mathbf{v} \tag{8.1}$$

where in a three-dimensional Cartesian space

$$\nabla = \frac{\partial}{\partial x}\mathbf{i} + \frac{\partial}{\partial y}\mathbf{j} + \frac{\partial}{\partial z}\mathbf{k} \tag{8.2}$$

$$\mathbf{v} = v_x\mathbf{i} + v_y\mathbf{j} + v_z\mathbf{k} \tag{8.3}$$

with $v_{x/y/z}$ being the local solution velocities in the x-, y- or z-direction $(\mathrm{m\,s^{-1}})$.

The characteristics of the fluid velocity depend on the design of the hydrodynamic cell and the flow pattern. The latter is said to be *laminar* when the solution flows smoothly and constantly in parallel layers such that the predominant velocity is that in the direction of the flow. Laminar flow conditions are desirable since accurate descriptions of the solution hydrodynamics are available. On the other hand, under *turbulent* flow conditions the solution motion is chaotic and the velocities in the directions perpendicular to that of the flow are significant. The transition between the laminar and turbulent regimes is defined in terms of the dimensionless Reynolds number, Re, that is proportional to the relative movement rate between the electrode and the solution, and to the electrode size, but inversely proportional to the kinematic viscosity of the solution. Thus, for low Re values the flow pattern is laminar and it transits to turbulent as Re increases. For example, in a tubular channel the laminar regime holds for $Re < 2300$.

From Eq. (8.1), the corresponding material balance equation establishes that

$$\frac{\partial c_j}{\partial t} = D_j \left(\frac{\partial^2 c_j}{\partial x^2} + \frac{\partial^2 c_j}{\partial y^2} + \frac{\partial^2 c_j}{\partial z^2} \right) - \left(v_x \frac{\partial c_j}{\partial x} + v_y \frac{\partial c_j}{\partial y} + v_z \frac{\partial c_j}{\partial z} \right) \tag{8.4}$$

In practice, the electrode and electrochemical cell as well as the experimental conditions are selected in order that transport by diffusion and convection dominates in one direction and Eq. (8.4) greatly simplifies. In the following sections we are going to consider two of the most popular hydrodynamic systems: the rotating disc electrode and the channel electrode. In both cases the mathematical problem can be reduced to a form analogous to that introduced in previous chapters for diffusion-only problems and then the same numerical strategies can be employed.

8.1. Rotating Disc Electrodes

The rotating disc electrode (RDE) is one of the most popular hydrodynamic electrodes due to its relatively easy fabrication, commercial availability and facile surface regeneration. Moreover, the corresponding simulation problem can be simplified to a single spatial dimension.

Fig. 8.1.   Schematic diagram of the rotating disc electrode with the corresponding flow profile.

As can be seen in Figure 8.1, the RDE consists of a disc of radius r_e embedded in an insulating rod that rotates at a constant frequency, f, in the electrolytic solution. As a result, the solution is pumped towards the disc surface and then thrown outwards. The Reynolds number in this case can be defined as

$$Re = \frac{f r_e^2}{v} \tag{8.5}$$

where f is the rotational speed (Hz), r_e the radius of the electrode and v the kinematic viscosity of the medium (m^2 s^{-1}) (e.g., $v \approx 10^{-6}$ m^2 s^{-1} in water at 25 °C). The flow pattern at the RDE under laminar conditions was studied by Karman [5] and Cochran [6] and it shows that the surface of the electrode is uniformly accessible, which greatly simplifies the theoretical treatment. Thus, considering the symmetry of the system and neglecting edge effects, the continuity equation for species j is reduced to one dimension, y, corresponding to the distance to the electrode surface in its normal direction:

$$\frac{\partial c_j}{\partial t} = D\frac{\partial^2 c_j}{\partial y^2} - v_y\frac{\partial c_j}{\partial y} \tag{8.6}$$

where the local fluid velocity in the y-direction is given by [7]

$$v_y = -Ly^2\left[1 - \frac{0.33333}{0.51023}y\left(\frac{2\pi f}{v}\right)^{1/2} + \frac{0.10265}{0.51023}y^2\left(\frac{2\pi f}{v}\right)\right.$$

$$- \frac{0.01265}{0.51023}y^3\left(\frac{2\pi f}{v}\right)^{3/2} - \frac{0.00283}{0.51023}y^4\left(\frac{2\pi f}{v}\right)^2$$

$$\left. + \frac{0.00179}{0.51023}y^5\left(\frac{2\pi f}{v}\right)^{5/2} - \frac{0.00045}{0.51023}y^6\left(\frac{2\pi f}{v}\right)^3 ...\right] \tag{8.7}$$

that near the electrode surface $(y \to 0)$ can be approximated as

$$v_y \approx -Ly^2 \tag{8.8}$$

where

$$L = 0.51023\,(2\pi f)^{3/2}\,v^{-1/2} \tag{8.9}$$

Note that in the case of more complicated mechanisms, the corresponding kinetic terms must be added in Eq. (8.6) as studied in Chapters 5 and 6.

The boundary value problem is given by

$$\left.\begin{array}{l} t = 0,\ y \geq 0 \\ t > 0,\ y \to \infty \end{array}\right\} \quad c_A = c_A^*,\quad c_B = 0 \tag{8.10}$$

$t > 0,\ y = 0$:

$$D\left(\frac{\partial c_A}{\partial y}\right)_{y=0} = k_{\mathrm{red}}c_A(0) - k_{\mathrm{ox}}c_B(0)$$

$$D\left(\frac{\partial c_B}{\partial y}\right)_{y=0} = -D\left(\frac{\partial c_A}{\partial y}\right)_{y=0} \tag{8.11}$$

The above mathematical problem can be written in dimensionless form by defining the dimensionless variables:

$$W = \left(\frac{L}{D}\right)^{1/3} y \tag{8.12}$$

$$T = \left(L^2 D\right)^{1/3} t \tag{8.13}$$

such that the continuity equations for species A and B are given by

$$\frac{\partial C_A}{\partial T} = \frac{\partial^2 C_A}{\partial W^2} + W^2 \frac{\partial C_A}{\partial W}$$

$$\frac{\partial C_B}{\partial T} = \frac{\partial^2 C_B}{\partial W^2} + W^2 \frac{\partial C_B}{\partial W} \tag{8.14}$$

where $C_j = c_j / c_A^*$. Note that with the new definition of the dimensionless time, T, the dimensionless scan rate in cyclic voltammetry is given by

$$\sigma = \frac{F\nu}{\mathcal{R}\mathcal{T}} \left(L^2 D\right)^{-1/3} \tag{8.15}$$

For the resolution of the problem the use of the Hale transformation [8] is convenient [9]:

$$Y = \frac{\int\limits_0^W \exp\left(-\frac{1}{3}W^3\right) \mathrm{d}W}{\int\limits_0^\infty \exp\left(-\frac{1}{3}W^3\right) \mathrm{d}W} \tag{8.16}$$

such that the value $Y = 0$ corresponds to the electrode surface and $Y = 1$ to an infinite distance from it, which enables us to cover the real space in its entirety with the simulation. Moreover, with the Hale transformation the differential equations simplify to

$$\frac{\partial C_A}{\partial T} = \frac{\exp\left(-\frac{2}{3}W^3\right)}{\left[\int\limits_0^\infty \exp\left(-\frac{1}{3}W^3\right) \mathrm{d}W\right]^2} \frac{\partial^2 C_A}{\partial Y^2}$$

$$\frac{\partial C_B}{\partial T} = \frac{\exp\left(-\frac{2}{3}W^3\right)}{\left[\int\limits_0^\infty \exp\left(-\frac{1}{3}W^3\right) \mathrm{d}W\right]^2} \frac{\partial^2 C_B}{\partial Y^2} \tag{8.17}$$

where

$$\left[\int\limits_0^\infty \exp\left(-\frac{1}{3}W^3\right) \mathrm{d}W\right]^2 = 1.65894 \tag{8.18}$$

and the boundary conditions are given by

$$
\left.\begin{array}{l}
T = 0,\ Y \geq 0 \\
T > 0,\ Y = 1
\end{array}\right\} \quad C_A = C_A^*, \quad C_B = 0 \tag{8.19}
$$

$T > 0,\ Y = 0$:

$$
\left(\frac{\partial C_A}{\partial Y}\right)_{Y=0} = K_{red}C_A(0) - K_{ox}C_B(0)
$$

$$
\left(\frac{\partial C_B}{\partial Y}\right)_{Y=0} = -\left(\frac{\partial C_A}{\partial Y}\right)_{Y=0} \tag{8.20}
$$

where

$$
K_{red/ox} = \sqrt{1.65894}\, k_{red/ox} \left(D^2 L\right)^{-1/3} \tag{8.21}
$$

Considering a uniform spatial grid with an interval ΔY, the spatial derivatives can be discretised as

$$
\frac{\partial^2 C}{\partial Y^2} \approx \frac{C_{i-1} - 2C_i + C_{i+1}}{\Delta Y^2}
$$

$$
\left(\frac{\partial C}{\partial Y}\right)_{Y=0} \approx \frac{C_1 - C_0}{\Delta Y} \tag{8.22}
$$

The above RDE problem has been solved successfully making use of the explicit method [9]. Within this approach, according to Eq. (8.17) the concentration of species j at the point in solution i at the timestep k can be calculated directly from the previous profile, $C_{j,i}^{k-1}$, from

$$
C_{j,i}^k = C_{j,i}^{k-1} + \Delta T \frac{\exp\left(-\frac{2}{3}W_i^3\right)}{1.65894}\left(\frac{C_{j,i-1}^{k-1} - 2C_{j,i}^{k-1} + C_{j,i+1}^{k-1}}{\Delta Y^2}\right) \tag{8.23}
$$

For this we first need to establish the correspondence between the Y- and W-coordinates in order to calculate the factor $\exp\left(-\frac{2}{3}W_i^3\right)$ for each Y_i value. This can be done by integration of the following ordinary differential equation:

$$
\frac{dW}{dY} = g(W) \tag{8.24}
$$

where

$$
g(W) = \sqrt{1.65894}\, \exp\left(\frac{1}{3}W^3\right) \tag{8.25}
$$

For example, we can make use of the fourth-order Runge–Kutta method such that

$$W_i = W_{i-1} + \frac{1}{6}\left(k_1 + 2k_2 + 2k_3 + k_4\right)$$

$$k_1 = \Delta Y\, g(W_{i-1})$$

$$k_2 = \Delta Y\, g\left(W_{i-1} + \frac{k_1}{2}\right)$$

$$k_3 = \Delta Y\, g\left(W_{i-1} + \frac{k_2}{2}\right) \tag{8.26}$$

$$k_4 = \Delta Y\, g(W_{i-1} + k_3)$$

where $W_0 = 0$ as corresponds to $Y = 0$.

Once the concentration profile has been obtained, the current response is calculated from

$$J_{\text{RDE}} = \frac{I\sqrt{1.65894}}{F A c_A^* \left(D^2 L\right)^{1/3}} = -\left(\frac{\partial C_A}{\partial Y}\right)_{Y=0} \tag{8.27}$$

The resolution of the problem with the backward implicit scheme can also be carried out in a simple way given that the mathematical problem is formally equivalent to the diffusion-only one studied in Chapters 2–4. Thus, the convection-diffusion differential equation in discrete form for species j at the point i in solution is given by

$$\frac{C_{j,i}^k - C_{j,i}^{k-1}}{\Delta T} = \frac{\exp\left(-\frac{2}{3}W_i^3\right)}{1.65894}\left(\frac{C_{j,i-1}^k - 2C_{j,i}^k + C_{j,i+1}^k}{\Delta Y^2}\right) \tag{8.28}$$

that can be written as

$$\alpha_{j,i} C_{j,i-1}^k + \beta_{j,i} C_{j,i}^k + \gamma_{j,i} C_{j,i+1}^k = \delta_{j,i}^k \tag{8.29}$$

where

$$\alpha_{j,i} = -\frac{\exp\left(-\frac{2}{3}W_i^3\right)}{1.65894}\frac{\Delta T}{\Delta Y^2}$$

$$\beta_{j,i} = 2\frac{\exp\left(-\frac{2}{3}W_i^3\right)}{1.65894}\frac{\Delta T}{\Delta Y^2} + 1$$

$$\gamma_{j,i} = -\frac{\exp\left(-\frac{2}{3}W_i^3\right)}{1.65894}\frac{\Delta T}{\Delta Y^2} \tag{8.30}$$

$$\delta_{j,i}^k = C_{j,i}^{k-1}$$

After redefining the coefficients α, β and γ due to the inference of convection, the form of the set of equations is equivalent to that studied in Chapters 3 and 4 and then it can be solved with the classical Thomas algorithm.[1]

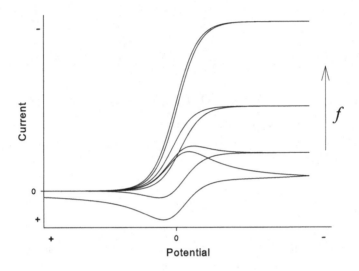

Fig. 8.2.   Influence of the rotational speed, f, on the cyclic voltammetry at the rotating disc electrode.

8.1.1.   *Testing Simulations: Specific tests*

Figure 8.2 shows the variation of the cyclic voltammograms of a reversible, one-electron reduction process at a rotating disc electrode when the rotational speed, f, is increased. The current increases with f given the faster supply of electroactive species by convection. Analogously to the case of microelectrodes, as the mass transport is more effective and the system approaches the steady state, the backward peak disappears and the diffusional shape of the forward one turns into a sigmoid given by

$$J_{\text{RDE,ss}} = -\frac{K_{\text{red}}\left(1 - \frac{c_B^*}{c_A^*}e^\theta\right)}{1 + K_{\text{red}}\left(1 + \frac{e^\theta}{d_B^{2/3}}\right)} \tag{8.31}$$

---

[1] Likewise, in the case of more complex mechanisms the methods described in Chapters 5 and 6 can be employed.

the steady-state (*plateau*) limiting current of which was derived by Levich under the first-order convection approximation [10, 11]:

$$|J_{\text{RDE,plateau}}| = 1 \tag{8.32}$$

In the case of fully reversible electron transfers, the expression (8.31) turns into

$$J_{\text{RDE,ss}} = -\frac{1 - \frac{c_B^*}{c_A^*}e^\theta}{1 + \frac{e^\theta}{d_B^{2/3}}} \tag{8.33}$$

and the corresponding reversible half-wave potential is given by

$$\theta_{1/2,\text{RDE,ss}} = \frac{2}{3}\ln(d_B) \tag{8.34}$$

8.2. Channel Electrodes

In the case of channel electrodes, the solution containing the electroactive species flows in a channel such as that shown in Figure 8.3 where a rectangular electrode of length x_e and width w is placed on the channel floor. The mass transport by convection can be controlled through the channel design, the electrode size and the flow rate. Moreover, this setup enables the incorporation of electrochemical measurements to flow systems as well as its use in spectroelectrochemistry and photoelectrochemistry [4].

Fig. 8.3. Schematic diagram of the channel electrode with the corresponding flow profile.

The cell design and flow rate are adjusted so that the flow pattern on the electrode is laminar and it can be described accurately for quantitative analysis. Under these conditions (that is, small Re values) and considering that the channel extends for enough upstream of the electrode for the flow to fully develop, then the solution velocity profile is parabolic and given by

$$v_x = v_0\left[1 - \frac{(h-y)^2}{h^2}\right], \quad v_y = 0, \; v_z = 0 \tag{8.35}$$

where v_0 is the solution velocity at the centre of the channel (m s^{-1}) and h is the half-height of the cell. Thus, velocity is maximum at the centre of the channel, as can be seen in Figure 8.3. For this system the Reynolds number can be defined as

$$Re = \frac{2hv_{\mathrm{m}}}{v} \tag{8.36}$$

where v_{m} is the mean flow velocity given by

$$v_{\mathrm{m}} = \frac{2}{3}v_0 \tag{8.37}$$

Unlike the RDE, convection and diffusion operate in different directions, the electrode surface is non-uniformly accessible and thus the associated mathematical problem is multidimensional. However, this can be simplified under appropriate experimental conditions. Thus, we can often assume that the flow velocity is high enough for the transport by diffusion in the direction of the flow (x-direction) to be negligible relative to convection. This is a valid approximation if the flow rate is fast and the electrode is large.[2] Moreover, edge effects can be neglected provided that the width of the band is much smaller than that of the channel ($w \ll d$), $h \ll d$ and $x_{\mathrm{e}} \ll w$.

From Eq. (8.4), and taking into account all the above considerations, the mass balance equation simplifies to

$$\frac{\partial c_j}{\partial t} = D\frac{\partial^2 c_j}{\partial y^2} - v_{\mathrm{x}}\frac{\partial c_j}{\partial x} \tag{8.38}$$

where we have considered that there is no migration or chemical reactions coupled to the electron transfer reaction and that the diffusion coefficients of the redox species A and B are equal. The corresponding boundary conditions are given by

$$\left.\begin{array}{l} t = 0, \ 0 \leq x \leq x_{\mathrm{e}}, \ 0 \leq y \leq 2h \\ t > 0, \ x < 0 \text{ (upstream of the electrode)} \end{array}\right\}$$

$$c_{\mathrm{A}} = c_{\mathrm{A}}^*, \ c_{\mathrm{B}} = 0 \tag{8.39}$$

---

[2] If microbands are employed and/or flow rates are slow, axial diffusion needs to be considered: $\frac{\partial c_j}{\partial t} = D\frac{\partial^2 c_j}{\partial x^2} + D\frac{\partial^2 c_j}{\partial y^2} - v_{\mathrm{x}}\frac{\partial c_j}{\partial x}$ [12].

$t > 0,\ 0 \le x \le x_e,\ y = 2h$:

$$\left(\frac{\partial c_A}{\partial y}\right)_{y=2h} = 0$$

$$\left(\frac{\partial c_B}{\partial y}\right)_{y=2h} = 0$$

(8.40)

$t > 0,\ 0 \le x \le x_e,\ y = 0$:

$$D\left(\frac{\partial c_A}{\partial y}\right)_{y=0} = k_{red}c_A(0) - k_{ox}c_B(0)$$

$$\left(\frac{\partial c_B}{\partial y}\right)_{y=0} = -\left(\frac{\partial c_A}{\partial y}\right)_{y=0}$$

(8.41)

By using the length of the band electrode, x_e, as the characteristic dimension and the Péclet number, Pe, that reflects the relative influence of the mass transport by convection and diffusion

$$Pe = \frac{v_m x_e}{D}$$

(8.42)

the following dimensionless variables and parameters can be defined:

$$X = \frac{x}{x_e},\ Y = \frac{y}{x_e},\ H = \frac{h}{x_e}$$

$$T = \frac{Dt}{x_e^2}$$

(8.43)

such that the dimensionless form of the convection-diffusion equations and the boundary value problem is

$$\frac{\partial C_A}{\partial T} = \frac{\partial^2 C_A}{\partial Y^2} - \frac{3}{2}Pe\left[1 - \frac{(H-Y)^2}{H^2}\right]\frac{\partial C_A}{\partial X}$$

$$\frac{\partial C_B}{\partial T} = \frac{\partial^2 C_B}{\partial Y^2} - \frac{3}{2}Pe\left[1 - \frac{(H-Y)^2}{H^2}\right]\frac{\partial C_B}{\partial X}$$

(8.44)

$$\left.\begin{array}{l} T = 0,\ 0 \le X \le 1,\ 0 \le Y \le 2H \\ T > 0,\ X < 0 \end{array}\right\}$$

$$C_A = 1,\ C_B = 0$$

(8.45)

$T > 0$, $0 \leq X \leq 1$, $Y = 2h/x_e$:

$$
\left(\frac{\partial C_A}{\partial Y} \right)_{Y=2h/x_e} = 0
$$

$$
\left(\frac{\partial C_B}{\partial Y} \right)_{Y=2h/x_e} = 0
$$

(8.46)

$T > 0$, $0 \leq X \leq 1$, $Y = 0$:

$$
\left(\frac{\partial C_A}{\partial Y} \right)_{Y=0} = K_{red} C_A(0) - K_{ox} C_B(0)
$$

$$
\left(\frac{\partial C_B}{\partial Y} \right)_{Y=0} = - \left(\frac{\partial C_A}{\partial Y} \right)_{Y=0}
$$

(8.47)

where $C_j = c_j/c_A^*$, $C_B^* = c_B^*/c_A^*$ and

$$
K_{red/ox} = \frac{k_{red/ox} x_e}{D}
$$

(8.48)

We can make use of the uniform, two-dimensional spatial grid shown in Figure 8.4 where the point $i = 0$ corresponds to a point upstream of the electrode where bulk concentrations of the electroactive species apply. The spatial derivatives can be approximated as

$$
\frac{\partial^2 C}{\partial Y^2} \approx \frac{C_{i,m-1} - 2C_{i,m} + C_{i,m+1}}{\Delta Y^2}
$$

$$
\frac{\partial C}{\partial X} \approx \frac{C_{i,m} - C_{i-1,m}}{\Delta X}
$$

$$
\left(\frac{\partial C}{\partial Y} \right)_{Y=0} \approx \frac{C_{i,1} - C_{i,0}}{\Delta Y}
$$

$$
\left(\frac{\partial C}{\partial Y} \right)_{Y=2h/x_e} \approx \frac{C_{i,N_Y-1} - C_{i,N_Y-2}}{\Delta Y}
$$

(8.49)

Note that the backward difference approximation has been employed for the first derivative with respect to X so that the problem can be solved easily by following a space-marching backward implicit method [13]. Thus, at a given timestep k the concentration profile in the y-direction for a fixed i value is obtained provided that the concentrations at $i - 1$ for all m are known. We start at $i = 1$ given that we know that the concentrations at $i = 0$ (for all m) are C_j^* and then we continue downstream up to $i = N_x$. This implies that we have to solve N_x problems for which the Thomas

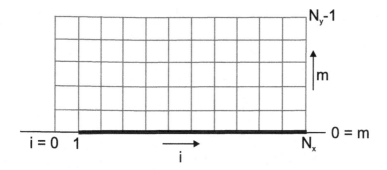

Fig. 8.4. Spatial grid for the numerical resolution of the channel electrode problem.

algorithm can be applied. Thus, with the backward implicit integration scheme the following set of linear equations needs to be solved for each i value from 1 to N_x:

$$\left(-\frac{\Delta T}{\Delta Y^2}\right)C_{j,i,m-1}^k + \left[\frac{2\Delta T}{\Delta Y^2} + \frac{3}{2}Pe\frac{m\Delta Y\Delta T}{\Delta XH^2}(2H - m\Delta Y) + 1\right]C_{j,i,m}^k$$

$$+ \left(-\frac{\Delta T}{\Delta Y^2}\right)C_{j,i,m+1}^k = C_{j,i-1,m}^{k-1} + \frac{3}{2}Pe\frac{m\Delta Y\Delta T}{\Delta XH^2}(2H - m\Delta Y)C_{j,i-1,m}^k$$

$$(8.50)$$

As in the case of the RDE, the problem for each i value is formally identical to that of a diffusion-only problem tackled in Chapters 3 and 4 and then the Thomas algorithm can be applied for its resolution. Once the full concentration profile has been calculated, that is, the concentrations for all the (i, m) points are known for the timestep k, the current response is calculated from

$$J_{\text{channel}} = \frac{I\,x_e}{FAD_A c_A^*} = -\int_0^1 \left(\frac{\partial C_A}{\partial Y}\right)_{Y=0} dX \qquad (8.51)$$

that can be evaluated with the trapezium method as described in Chapter 9. Under steady-state conditions, the following analytical equation describes the current-potential curve [14, 15]:

$$J_{\text{chan,ss}} = -0.925\left(\frac{x_e^2 v_f}{D_A h^2 d}\right)^{1/3}$$

$$\frac{1}{1 + d_B^{-2/3}e^\theta}\left[1 - 2u + 2u^2 \ln\left(\frac{1+u}{u}\right)\right]$$

$$(8.52)$$

where v_f is the volume flow rate (m^3 s^{-1})

$$v_f = \frac{4}{3}v_0 hd \tag{8.53}$$

and

$$u = \frac{0.6783 D_B^{2/3} \left(\frac{3v_f}{4h^2 dx_e}\right)^{1/3}}{k_{ox} + d_B^{2/3} k_{red}} \tag{8.54}$$

such that the mass-transport-limited current ($e^\theta \to 0$) is given by the Levich equation

$$|J_{chan,plateau}| = 0.925 \left(\frac{x_e^2 v_f}{D_A h^2 d}\right)^{1/3} \tag{8.55}$$

References

[1] J. K. Novev and R. G. Compton, Natural convection effects in electrochemical systems, *Current Opinion in Electrochemistry*. **7**, 118–129, (2018).

[2] A. Frumkin, L. Nekrasov, B. Levich, and J. Ivanov, Die Anwendung der rotierenden Scheibenelektrode mit einem Ring zur Untersuchung von Zwischenprodukten elektrochemischer Reaktionen, *J. Electroanal. Chem.* **1**, 84–90, (1959).

[3] J. Yamada and H. Matsuda, Limiting diffusion currents in hydrodynamic voltammetry: III. Wall jet electrodes, *J. Electroanal. Chem.* **44**, 189–198, (1973).

[4] J. A. Cooper and R. G. Compton, Channel electrodes. A review, *Electroanal.* **10**, 141–155, (1998).

[5] T. von Karman, Uber laminare und turbulente Reibung, *Z. Angew. Math. Mech.* **1**, 233–252, (1921).

[6] W. G. Cochran. The flow due to a rotating disc. In *Mathematical Proceedings of the Cambridge Philosophical Society*, vol. 30, pp. 365–375. Cambridge University Press, Cambridge, (1934).

[7] S. W. Feldberg, M. L. Bowers, and F. C. Anson, Hopscotch-finite-difference simulation of the rotating ring-disc electrode, *J. Electroanal. Chem.* **215**, 11–28, (1986).

[8] J. M. Hale, Transients in convective systems: I. Theory of galvanostatic, and galvanostatic with current reversal transients, at a rotating disc electrode, *J. Electroanal. Chem.* **6**, 187–197, (1963).

[9] R. G. Compton, M. E. Laing, D. Mason, R. J. Northing, and P. R. Unwin, Rotating disc electrodes: The theory of chronoamperometry and its use in mechanistic investigations, *R. Soc. Lond. Proc. Ser. A Math. Phys. Eng. Sci.* **418**, 113–154, (1988).

[10] V. G. Levich, *Physicochemical Hydrodynamics.* (Prentice-Hall, Englewood Cliffs, NJ, 1962).

[11] R. G. Compton and C. E. Banks, *Understanding Voltammetry, 3rd ed.* (World Scientific Press, Singapore, UK, 2018).

[12] R. G. Compton, A. C. Fisher, R. G. Wellington, P. J. Dobson, and P. A. Leigh, Hydrodynamic voltammetry with microelectrodes: channel microband electrodes - Theory and experiment, *J. Phys. Chem.* **97**, 10410–10415, (1993).

[13] R. G. Compton, M. B. G. Pilkington, and G. M. Stearn, Mass transport in channel electrodes, *J. Chem. Soc. Faraday Trans. 1.* **84**, 2155–2171, (1988).

[14] L. N. Klatt and W. J. Blaedel, Quasi-reversible and irreversible charge transfer at the tubular electrode, *Anal. Chem.* **39**, 1065–1072, (1967).

[15] N. V. Rees, J. A. Alden, R. A. W. Dryfe, B. A. Coles, and R. G. Compton, Voltammetry under high mass transport conditions. The high speed channel electrode and heterogeneous kinetics, *J. Phys. Chem.* **99**, 14813–14818, (1995).

Chapter 9

Two-Dimensional Systems: Microdisc Electrodes

9.1. Microdisc Electrodes: The Model

In this chapter, we consider the cyclic voltammetry of a simple one-electron reduction at a microdisc electrode. The change from modelling a macroscale electrode to modelling a microscale one corresponds to a significant increase in mathematical complexity. Therefore most of the additional concepts developed in the preceding chapters such as unequal diffusion coefficients, homogeneous reactions and other experimental techniques are not included in the text and we focus on the relatively simple problem of simulating cyclic voltammetry of a one-electron reduction process. As such, this chapter may be successfully followed without first reading any of Chapters 5–8. It is expected, however, that once the reader fully understands the techniques for simulating a microelectrode, the same additional concepts covered in those chapters may easily be extended for use with the microdisc model.

A microdisc electrode is a micron-scale flat conducting disc of radius r_e that is embedded in an insulating surface, with the disc surface flush with that of the insulator. It is assumed that electron transfer takes place only on the surface of the disc and that the supporting surface is completely electroinactive under the conditions of the experiment. These electrodes are widely employed in electrochemical measurements since they offer the advantages of microelectrodes (reduced ohmic drop and capacitive effects, miniaturisation of electrochemical devices) and are easy to fabricate and clean for surface regeneration. In Chapter 2, we considered a disc-shaped electrode of size on the order of 1 mm. In that case we could approximate the system as one-dimensional because the electrode was large in comparison to the thickness of the diffusion layer, such that the current was essentially uniform across the entire electrode surface. Due to the small size of the microdisc, this approximation is no longer valid so we must work in terms of a three-dimensional coordinate system. While the microdisc can be considered in terms of Cartesian coordinates (x, y, z), with the electrode surface in the (x, y)-plane, the geometry lends itself well to treatment in terms of cylindrical polar coordinates (r, z, φ) as illustrated in Figure 9.1.

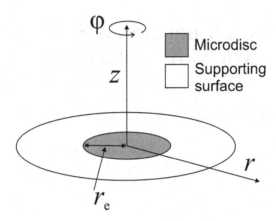

Fig. 9.1.   The (r, z, φ) cylindrical polar coordinate system used to model a microdisc electrode. The disc radius is r_e.

In this system, an axis passes through the centre of the disc perpendicular to the plane of its surface. r is the radial distance from the axis, z is the perpendicular distance from the surface and φ is the angle around the axis. The origin is at the centre of the disc.

While it is true that the current at the electrode surface will vary with radial coordinate, r, it can be seen that there will be no variation with angle φ,

$$\frac{\partial c}{\partial \varphi} = 0 \tag{9.1}$$

as the system is cylindrically symmetrical. The problem is therefore reduced to two dimensions, r and z, and we need only model a single (r, z)-plane as illustrated in Figure 9.2; integration across all φ achieves the full three-dimensional result.

9.1.1. *Diffusion*

Fick's second law predicts how diffusion causes the concentration to change with time. In general, this may be stated as

$$\frac{\partial c}{\partial t} = D\nabla^2 c \tag{9.2}$$

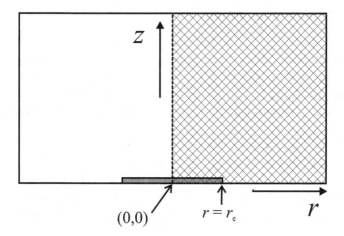

Fig. 9.2. The two-dimensional (r, z) slice of the microdisc model. As the system has mirror symmetry in the line $z = 0$, only one half, indicated by cross-hatching, needs to be considered.

where $\nabla^2 c$ is the second derivative of the concentration in space in the relevant coordinate system. In three-dimensional (x, y, z) Cartesian coordinates this is

$$\frac{\partial c}{\partial t} = D \left(\frac{\partial^2 c}{\partial x^2} + \frac{\partial^2 c}{\partial y^2} + \frac{\partial^2 c}{\partial z^2} \right) \tag{9.3}$$

Transforming this equation into cylindrical polar (r, z, φ) coordinates gives

$$\frac{\partial c}{\partial t} = D \left(\frac{\partial^2 c}{\partial r^2} + \frac{1}{r} \frac{\partial c}{\partial r} + \frac{1}{r^2} \frac{\partial^2 c}{\partial \varphi^2} + \frac{\partial^2 c}{\partial z^2} \right) \tag{9.4}$$

However, as the model microdisc system is cylindrically symmetrical, it is true that:

$$\frac{\partial c}{\partial \varphi} = \frac{\partial^2 c}{\partial \varphi^2} = 0 \tag{9.5}$$

and so to model diffusion in the (r, z) slice depicted in Figure 9.2, we use the equation:

$$\frac{\partial c}{\partial t} = D \left(\frac{\partial^2 c}{\partial r^2} + \frac{1}{r} \frac{\partial c}{\partial r} + \frac{\partial^2 c}{\partial z^2} \right) \tag{9.6}$$

9.1.2. *Boundary conditions*

In a one-dimensional system, the simulation space is a line, so the space is bounded by two singular points: the electrode surface and the bulk concentration boundary. In this two-dimensional system, the simulation space is a rectangular region of a flat plane, so four boundary lines, each with their own boundary conditions, are required. These boundary lines are $r = 0$, $z = 0$, $r = r_{max}$ and $z = z_{max}$.

We assume that the microdisc is embedded in a planar supporting surface that is infinite in extent and that the solution is also infinite in extent. Far away from the electroactive surface where the solution is unperturbed by the electrode reactions, the concentration of all species is always equal to their bulk value, c^*. Therefore we have the boundary conditions:

$$c(r = r_{max}, z) = c^* \qquad (9.7)$$

$$c(r, z = z_{max}) = c^* \qquad (9.8)$$

where the semi-infinite extent of the simulation space (see Chapter 2) is given by

$$r_{max} = r_e + 6\sqrt{Dt_{max}} \qquad (9.9)$$

$$z_{max} = 6\sqrt{Dt_{max}} \qquad (9.10)$$

For the electrode plane ($z = 0$), two separate boundary conditions are required as the plane is composed of two different materials: the electroactive microdisc and the insulating supporting surface. As the microdisc surface is electroactive, a potential-dependent boundary condition is applied; the Nernst equation, Butler–Volmer or Marcus–Hush models may be used as appropriate:

$$c(r \le r_e, z = 0) = f(E) \qquad (9.11)$$

The insulating supporting surface is a solid boundary so there can be no flux of solution-phase material across it, i.e., species A and B cannot pass through it. The boundary condition therefore specifies that the flux at the boundary, which is proportional to the first derivative of concentration, is zero:

$$\left(\frac{\partial c(r > r_e)}{\partial z}\right)_{z=0} = 0 \qquad (9.12)$$

The final spatial boundary is the cylindrical axis, $r = 0$. As illustrated in Figure 9.2, this axis is a line of mirror symmetry. Consequently, for any

Fig. 9.3. The two-dimensional simulation space for a microdisc electrode.

given value of z, the concentration at the two points on either side of the boundary is exactly equal, so the concentration gradient at the boundary must necessarily be zero. The boundary condition is therefore

$$\left(\frac{\partial c}{\partial r}\right)_{r=0} = 0 \tag{9.13}$$

In addition to the spatial boundary conditions, solution of the model also requires a set of initial conditions, i.e., a specification of the initial state of the system. As with the one-dimensional system, we assume that the concentration of each species is initially uniform and equal to its bulk value across all space:

$$c(r, z, t = 0) = c^* \tag{9.14}$$

The mathematical model of the microdisc system and its boundary conditions are summarised in Figure 9.3.

9.1.3. Current

For a one-dimensional macroelectrode, the current, I, is calculated as

$$I = FAj_{0,A} \tag{9.15}$$

where A is the total electrode area and $j_{0,A}$ is the flux of electroactive species at the electrode surface. This assumes uniform flux across the whole surface which is valid for a macroelectrode; however, for a microdisc electrode, the flux is obviously not uniform across the surface; it varies with r.

The flux is therefore calculated at every point across the radius of the disc (i.e., a line) and integrated across all angles, φ (which simply scales the result by a factor of 2π), to give the total flux for the whole electrode. The real current is therefore

$$I = 2\pi F j'_{0,A} \tag{9.16}$$

where $j'_{0,A}$ is the flux across the radius, which is calculated as:

$$j'_{0,A} = \int_0^{r_e} j_0 r \, dr = -D \int_0^{r_e} \left(\frac{\partial c_A}{\partial z} \right)_{z=0} r \, dr \tag{9.17}$$

where the factor of r arises from integration about angle φ.

9.1.4. *Normalisation*

The normalisation scheme used for the microdisc model is similar to that used for the one-dimensional model (Section 2.5). For the two-dimensional model there are two spatial coordinates, each of which is normalised against the disc radius, r_e:

$$R = \frac{r}{r_e} \tag{9.18}$$

$$Z = \frac{z}{r_e} \tag{9.19}$$

In these transformed coordinates, the radius of the microdisc is 1. The definitions of T and C are unchanged so that the dimensionless Fick's second law in this space is

$$\frac{\partial C}{\partial T} = \frac{\partial^2 C}{\partial R^2} + \frac{1}{R} \frac{\partial C}{\partial R} + \frac{\partial^2 C}{\partial Z^2} \tag{9.20}$$

Transformations of the boundary conditions are trivial. The real current, I, is calculated from

$$I = -2\pi F D c_A^* r_e \int_0^1 \frac{\partial C_A}{\partial Z} R \, dR \tag{9.21}$$

9.2. Numerical Solution

In the numerical solution of the one-dimensional system, we used a three-point central differencing scheme such that the first and second spatial derivatives of C at point i (for a grid with non-uniform spacing) were

$$\frac{dC}{dX} = \frac{C_{i+1} - C_{i-1}}{\Delta X_+ + \Delta X_-} \tag{9.22}$$

$$\frac{d^2C}{dX^2} = \left(\frac{C_{i+1} - C_i}{\Delta X_+} - \frac{C_i - C_{i-1}}{\Delta X_-} \right) \frac{2}{\Delta X_+ + \Delta X_-} \tag{9.23}$$

9.2.1. *Finite differences in two dimensions*

For one-dimensional simulations, the space was discretised, transforming it into an array of discrete points. In order to simulate a microdisc, the discretisation process transforms the continuous two-dimensional space into a two-dimensional (R, Z) grid of discrete spatial points. As with the one-dimensional case, the simulation efficiency may be greatly increased if we employ an expanding spatial grid with a higher density of points in areas where we expect greater concentration gradients and at finite spatial boundaries [1, 2]. The electrode surface requires a higher point density than does the bulk solution and therefore the grid should expand in the Z-direction (perpendicular to the electrode surface). Here we use the same expanding spatial mesh as was used for X in the one-dimensional case which is efficient but by no means the only sensible scheme available. The grid is defined by

$$Z_0 = 0 \tag{9.24}$$

$$Z_1 = h \tag{9.25}$$

$$Z_j = Z_{j-1} + h\omega^{(j-1)} \tag{9.26}$$

where h is the size of the first spatial increment and ω is the space expansion coefficient ($\omega > 1$). Decreasing the values of h and ω leads to an increase in the number of spatial points and therefore typically leads to increased accuracy at the expense of increased simulation runtime. As always, a convergence study is required to determine the optimum values of ω and h. A reasonable set of values[1] to use as a starting point is $\omega = 1.1$, $h = 10^{-4}$.

---

[1] Note that it is possible to specify independent values of ω and h for each direction and indeed this may be beneficial under certain circumstances; however, for simplicity here we use the same value of each for both the Z- and R-grids.

In the R-direction there are two finite boundaries that must be considered, namely the $R = 0$ axis, and the singularity where the outer edge of the electrode meets the insulating surface, at $R = 1$. The spatial grid must therefore expand outwards from these two points, giving a high density of points at both edges of the electrode and a lower density in the middle of it [1]:

$$0 < R \leq \frac{1}{2} \qquad R_i = R_{i-1} + h\omega^{(i-1)} \qquad (9.27)$$

$$\frac{1}{2} < R < 1 \qquad R_i = R_{i-1} + h\omega^{(n_e - i)} \qquad (9.28)$$

$$1 \leq R \leq R_{\max} \qquad R_i = 1 + h\omega^{(i - n_e - 1)} \qquad (9.29)$$

where n_e is the space point such that $R_{n_e} = 1$. Computationally, this may be achieved by determining the grid points for $0 < R \leq \frac{1}{2}$ and then mirroring them for $\frac{1}{2}$ to 1. This will mean that the expansion will be slightly uneven around $R = \frac{1}{2}$ but this will not adversely affect the simulation. An example of a two-dimensional grid of spatial points generated in this manner is shown in Figure 9.4. The set of values of the concentration at each spatial point is stored in a two-dimensional array in the computer's memory, which we will refer to as the *concentration grid*.

We apply the same three-point central differencing scheme as was used in the one-dimensional case to this two-dimensional space, which gives the

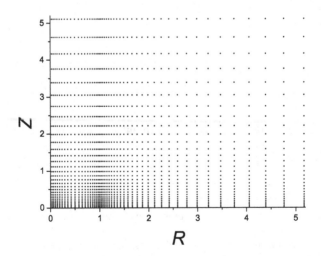

Fig. 9.4. An expanding spatial grid used for microdisc simulations.

form of the derivatives for the concentration at point (i, j) as

$$\frac{dC}{dR} = \frac{C_{i+1,j} - C_{i-1,j}}{\Delta R_+ + \Delta R_-} \tag{9.30}$$

$$\frac{dC}{dZ} = \frac{C_{i,j+1} - C_{i,j-1}}{\Delta Z_+ + \Delta Z_-} \tag{9.31}$$

$$\frac{d^2C}{dR^2} = \left(\frac{C_{i+1,j} - C_{i,j}}{\Delta R_+} - \frac{C_{i,j} - C_{i-1,j}}{\Delta R_-} \right) \frac{2}{\Delta R_+ + \Delta R_-} \tag{9.32}$$

$$\frac{d^2C}{dZ^2} = \left(\frac{C_{i,j+1} - C_{i,j}}{\Delta Z_+} - \frac{C_{i,j} - C_{i,j-1}}{\Delta Z_-} \right) \frac{2}{\Delta Z_+ + \Delta Z_-} \tag{9.33}$$

where i is the index of a discrete point in the R-direction, j is the index of a point in the Z-direction and

$$\Delta R_- = R_i - R_{i-1} \tag{9.34}$$

$$\Delta R_+ = R_{i+1} - R_i \tag{9.35}$$

$$\Delta Z_- = Z_j - Z_{j-1} \tag{9.36}$$

$$\Delta Z_+ = Z_{j+1} - Z_j \tag{9.37}$$

In Chapter 3, we chose to use an implicit method of solution, as opposed to an explicit one for reasons of stability and simulation efficiency (despite the greater complexity of the implicit method). The implicit discretisation of Eq. (9.20) is [3]

$$\frac{C_{i,j}^k - C_{i,j}^{k-1}}{\Delta T} = \left(\frac{C_{i+1,j}^k - C_{i,j}^k}{\Delta R_+} - \frac{C_{i,j}^k - C_{i-1,j}^k}{\Delta R_-} \right) \frac{2}{\Delta R_+ + \Delta R_-}$$

$$+ \left(\frac{C_{i,j+1}^k - C_{i,j}^k}{\Delta Z_+} - \frac{C_{i,j}^k - C_{i,j-1}^k}{\Delta Z_-} \right) \frac{2}{\Delta Z_+ + \Delta Z_-}$$

$$+ \frac{1}{R_i} \left(\frac{C_{i+1,j}^k - C_{i-1,j}^k}{\Delta R_+ + \Delta R_-} \right) \tag{9.38}$$

This is an equation of the form

$$C_{i,j}^{k-1} = \lambda_1 C_{i-1,j}^k + \lambda_2 C_{i,j}^k + \lambda_3 C_{i+1,j}^k + \lambda_4 C_{i,j-1}^k + \lambda_5 C_{i,j+1}^k \tag{9.39}$$

where the λs are constant coefficients. This equation contains only one known and five unknowns. The matrix required to solve this linear system is no longer tridiagonal and so cannot be solved using the Thomas algorithm. The generalised n-diagonal version of the Thomas algorithm can be used (see Chapter 5); however, this may not be numerically efficient and another solution is found in the alternating direction implicit (ADI) method.

9.2.2. The ADI method

In the ADI method [4–6], the explicit and implicit methods are combined. The timesteps, ΔT, are divided into two half-timesteps, $\Delta T/2$. For the first of these half-timesteps, $T_{k-\frac{1}{2}}$, the derivatives along the Z-coordinate are discretised implicitly, while the derivatives along the R-coordinate are discretised explicitly, and vice versa for the second half-timestep, T_k. Thus at the end of each timestep, each coordinate has received an equal share of implicit and explicit discretisation. This method is found to minimise error and maintain ΔT stability [7].

To solve for a given point, $C_{i,j}^{k-\frac{1}{2}}$, in the first half-timestep, there are therefore three unknowns: $C_{i,j-1}^{k-\frac{1}{2}}$, $C_{i,j}^{k-\frac{1}{2}}$ and $C_{i,j+1}^{k-\frac{1}{2}}$. All of these unknowns lie at the same value of R, i.e., they are all in the same column; consequently, we can solve for the whole concentration grid using the tridiagonal Thomas algorithm if we consider the spatial grid to be composed of a series of columns, each of which may be solved independently. Each column may be thought of as being equivalent to a one-dimensional concentration array, the only difference being two additional explicit terms at each point. We call the first half-timestep the Z-implicit sweep because the solution sweeps across the simulation space, solving one column at a time as depicted in Figure 9.5(a). In the second half-timestep, the R-implicit sweep, the three unknowns are $C_{i-1,j}^{k}$, $C_{i,j}^{k}$ and $C_{i+1,j}^{k}$ which all lie at the same value of Z and thus are all in the same row. In this case the solution sweeps up the space solving one row at a time as depicted in Figure 9.5(b).

As each row or column within a sweep may be solved independently of the others, there is no requirement to solve them in order; while a typical iterative approach would solve the columns in the Z-implicit sweep in the order R_0, R_1, R_2, etc., they may equally be solved in the order R_{49}, R_{26}, R_{37}, etc. An important consequence is that within a given sweep, many rows/columns can be solved in parallel if the machine on which the simulation is running has multiple processor cores. This can lead to a reduction in simulation runtime by a factor proportional to the number of available cores. This is discussed further in Subsection 9.3.1.

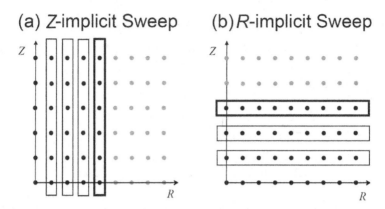

Fig. 9.5. Z-implicit and R-implicit solution sweeps. In the current sweep, thin-lined columns are already solved for, and the thick-lined columns are those currently being solved.

9.2.2.1. *Z-sweep discretisation*

For the Z-sweep, at each point in space we solve:

$$\frac{C_{i,j}^{k-\frac{1}{2}} - C_{i,j}^{k-1}}{\frac{1}{2}\Delta T} = \nabla^2 C(Z(T_{k-\frac{1}{2}}), R(T_{k-1})) \qquad (9.40)$$

i.e., using values from the previous half-timestep $(k-1)$ in the R-direction (explicit) and from the current timestep $(k-\frac{1}{2})$ in the Z-direction (implicit).

The discretisation for each of the terms appearing in Eq. (9.20) for the Z-sweep is as follows:

$$\frac{\partial^2 C}{\partial R^2} = \frac{2}{\Delta R_+ + \Delta R_-}\left(\frac{C_{i+1,j}^{k-1} - C_{i,j}^{k-1}}{\Delta R_+} - \frac{C_{i,j}^{k-1} - C_{i-1,j}^{k-1}}{\Delta R_-}\right) \qquad (9.41)$$

$$\frac{\partial^2 C}{\partial Z^2} = \frac{2}{\Delta Z_+ + \Delta Z_-}\left(\frac{C_{i,j+1}^{k-\frac{1}{2}} - C_{i,j}^{k-\frac{1}{2}}}{\Delta Z_+} - \frac{C_{i,j}^{k-\frac{1}{2}} - C_{i,j-1}^{k-\frac{1}{2}}}{\Delta Z_-}\right) \qquad (9.42)$$

$$\frac{1}{R}\frac{\partial C}{\partial R} = \frac{1}{R_i}\left(\frac{C_{i+1,j}^{k-1} - C_{i-1,j}^{k-1}}{\Delta R_+ + \Delta R_-}\right) \qquad (9.43)$$

Therefore at each space point we have an equation of the form

$$\alpha C_{i,j-1}^{k-\frac{1}{2}} + \beta C_{i,j}^{k-\frac{1}{2}} + \gamma C_{i,j+1}^{k-\frac{1}{2}} = \delta^{k-\frac{1}{2}} \qquad (9.44)$$

where

$$\delta^{k-\frac{1}{2}} = \lambda_1 C_{i-1,j}^{k-1} + \lambda_2 C_{i,j}^{k-1} + \lambda_3 C_{i+1,j}^{k-1} \tag{9.45}$$

The coefficients are thus

$$\alpha = \frac{2}{\Delta Z_+ + \Delta Z_-}\left(\frac{1}{\Delta Z_-}\right) \tag{9.46}$$

$$\beta = \frac{2}{\Delta Z_+ + \Delta Z_-}\left(-\frac{1}{\Delta Z_+} - \frac{1}{\Delta Z_-}\right) - \frac{2}{\Delta T} \tag{9.47}$$

$$\gamma = \frac{2}{\Delta Z_+ + \Delta Z_-}\left(\frac{1}{\Delta Z_+}\right) \tag{9.48}$$

$$\lambda_1 = -\frac{1}{\Delta R_+ + \Delta R_-}\left(\frac{2}{\Delta R_-} - \frac{1}{R_i}\right) \tag{9.49}$$

$$\lambda_2 = -\frac{2}{\Delta R_+ + \Delta R_-}\left(-\frac{1}{\Delta R_+} - \frac{1}{\Delta R_-}\right) - \frac{2}{\Delta T} \tag{9.50}$$

$$\lambda_3 = -\frac{1}{\Delta R_+ + \Delta R_-}\left(\frac{2}{\Delta R_+} + \frac{1}{R_i}\right) \tag{9.51}$$

9.2.2.2. *R-sweep discretisation*

For the R-sweep, at each point in space we solve

$$\frac{C_{i,j}^k - C_{i,j}^{k-\frac{1}{2}}}{\frac{1}{2}\Delta T} = \nabla^2 C(Z(T_{k-\frac{1}{2}}), R(T_k)) \tag{9.52}$$

The discretisation for each of the terms appearing in Eq. (9.20) for the R-sweep is as follows:

$$\frac{\partial^2 C}{\partial R^2} = \frac{2}{\Delta R_+ + \Delta R_-}\left(\frac{C_{i+1,j}^k - C_{i,j}^k}{\Delta R_+} - \frac{C_{i,j}^k - C_{i-1,j}^k}{\Delta R_-}\right) \tag{9.53}$$

$$\frac{\partial^2 C}{\partial Z^2} = \frac{2}{\Delta Z_+ + \Delta Z_-}\left(\frac{C_{i,j+1}^{k-\frac{1}{2}} - C_{i,j}^{k-\frac{1}{2}}}{\Delta Z_+} - \frac{C_{i,j}^{k-\frac{1}{2}} - C_{i,j-1}^{k-\frac{1}{2}}}{\Delta Z_-}\right) \tag{9.54}$$

$$\frac{1}{R}\frac{\partial C}{\partial R} = \frac{1}{R_i}\left(\frac{C_{i+1,j}^k - C_{i-1,j}^k}{\Delta R_+ + \Delta R_-}\right) \tag{9.55}$$

Therefore at each space point we have an equation of the form

$$\alpha C_{i-1,j}^k + \beta C_{i,j}^k + \gamma C_{i+1,j}^k = \delta^k \tag{9.56}$$

where

$$\delta^k = \lambda_1 C_{i,j-1}^{k-\frac{1}{2}} + \lambda_2 C_{i,j}^{k-\frac{1}{2}} + \lambda_3 C_{i,j+1}^{k-\frac{1}{2}} \tag{9.57}$$

The coefficients are thus

$$\alpha = \frac{1}{\Delta R_+ + \Delta R_-} \left(\frac{2}{\Delta R_-} - \frac{1}{R_i} \right) \tag{9.58}$$

$$\beta = \frac{2}{\Delta R_+ + \Delta R_-} \left(-\frac{1}{\Delta R_+} - \frac{1}{\Delta R_-} \right) - \frac{2}{\Delta T} \tag{9.59}$$

$$\gamma = \frac{1}{\Delta R_+ + \Delta R_-} \left(\frac{2}{\Delta R_+} + \frac{1}{R_i} \right) \tag{9.60}$$

$$\lambda_1 = -\frac{2}{\Delta Z_+ + \Delta Z_-} \left(\frac{1}{\Delta Z_-} \right) \tag{9.61}$$

$$\lambda_2 = -\frac{2}{\Delta Z_+ + \Delta Z_-} \left(-\frac{1}{\Delta Z_+} - \frac{1}{\Delta Z_-} \right) - \frac{2}{\Delta T} \tag{9.62}$$

$$\lambda_3 = -\frac{2}{\Delta Z_+ + \Delta Z_-} \left(\frac{1}{\Delta Z_+} \right) \tag{9.63}$$

9.2.3. Boundary conditions

In normalised coordinates, the initial and boundary conditions are as follows:

Initial conditions:	$T = 0$	$C = 1$
Electrode surface:	$Z = 0,\ R \leq 1$:	$C = f(\theta)$
Supporting surface:	$Z = 0,\ R > 1$:	$\frac{\partial C}{\partial Z} = 0$
Axial symmetry:	$R = 0$	$\frac{\partial C}{\partial R} = 0$
Bulk solution:	$R \to \infty,\ Z \to \infty$:	$C = 1$

Here we have given a general form for the electrode surface boundary condition, i.e., $C = f(\theta)$; we can choose to simulate a particular type of experiment with a given kinetic model by specifying the form of this function. In this case we aim to simulate the cyclic voltammetry of a one-electron reduction using Nernstian equilibrium, as with the macroelectrode in Chapters 2 and 3. Therefore this function takes the form

$$C(R \leq 1, Z = 0) = \frac{1}{1 + e^{-\theta}} \tag{9.64}$$

The initial conditions are satisfied by setting the value of every point in the concentration grid to 1 before the simulation begins. We must now consider how the rest of these conditions are represented in discrete form when they are treated implicitly (boundary perpendicular to the implicit direction) and explicitly (boundary parallel to the implicit direction).

9.2.3.1. *Implicit terms*

The electrode surface boundary conditions are only relevant in the Z-sweep as that sweep considers diffusion perpendicular to the surface. If we suppose Nernstian equilibrium, then the electrode boundary conditions are met by using the following coefficients for the first point, $j = 0$, of every column $n \leq n_e$ in the Z-sweep:

$$\alpha_0 = 0, \qquad \beta_0 = 1, \qquad \gamma_0 = 0, \qquad \delta_0^{(k)} = \frac{1}{1 + e^{-\theta^{(k)}}} \tag{9.65}$$

The boundary condition for the insulating supporting surface is also only relevant for the Z-sweep. In discrete form, the condition $\left(\frac{\partial C}{\partial Z}\right)_{Z=0} = 0$ is represented as

$$C_{i,1}^{k-\frac{1}{2}} - C_{i,0}^{k-\frac{1}{2}} = 0 \tag{9.66}$$

This is achieved by using the following coefficients for the first point of every column $n > n_e$ in the Z-sweep

$$\alpha_0 = 0, \qquad \beta_0 = -1, \qquad \gamma_0 = 1, \qquad \delta_0 = 0 \tag{9.67}$$

The axial symmetry boundary is a no-flux boundary at $R = 0$ and is handled in the same way as the insulating surface boundary; for every row in the R-sweep, the first point, $i = 0$, uses the coefficients given by Eq. (9.67).

The two bulk concentration boundaries are handled in the same manner as in the one-dimensional model, by setting the value of the concentration at the boundary to 1. For the Z-sweep, the last point in each column, $j = m - 1$, uses the following coefficients:

$$\alpha_{m-1} = 0, \qquad \beta_{m-1} = 1, \qquad \gamma_{m-1} = 0, \qquad \delta_{m-1} = 1 \tag{9.68}$$

such that $C_{i,m-1}^{k-\frac{1}{2}} = 1$ for all i. For the R-sweep, the same coefficients are used at the last point in each row, $i = n - 1$, such that $C_{n-1,j}^{k} = 1$ for all j. Note that since the bulk boundaries are sufficiently far away from the electrode surface that there can be essentially no perturbation of the concentration there, the concentration gradient at the boundary is zero, so a condition of no flux may be used as an alternative to a condition of constant concentration.

9.2.3.2. *Explicit terms*

For the simple one-electron reduction that we study here, it is possible to completely omit the boundary columns and rows without adversely affecting the simulation results as long as a reasonably dense spatial grid is employed. For the Z-sweep, we may simply ignore the first and last columns ($i = 0$ and $i = n - 1$ respectively), and for the R-sweep, the first and last rows ($j = 0$ and $j = m - 1$ respectively). This removes the need for extra code to deal with the altered explicit (λ) terms that are necessary when considering these boundary rows and columns. However, for simulations of more complicated situations, such as homogeneous kinetics, this simplification may not be appropriate when considering the axial symmetry boundary, therefore we provide implementation details here.

If we examine the axial symmetry boundary ($R = 0$) in the Z-sweep, we can see that if we consider a three-point approximation of $\frac{\partial C}{\partial R}$ and $\frac{\partial^2 C}{\partial R^2}$ (the explicit terms), the equation will involve concentration terms with an i index of -1, which is outside of our concentration grid. This is not a significant problem since this boundary is a line of mirror symmetry and therefore the value of $C_{-1,j}$ is always identical to that of $C_{1,j}$. However there is a further problem at this particular boundary due to the presence of the $1/R$ term since $1/R \to \infty$ as $R \to 0$. This situation may be resolved by considering the Maclaurin expansion [2, 8, 9] of $\frac{\partial C}{\partial R}$ at $R = 0$ which gives the relation

$$\frac{1}{R}\left(\frac{\partial C}{\partial R}\right)_{R=0} \approx \left(\frac{\partial^2 C}{\partial R^2}\right)_{R=0} \tag{9.69}$$

This can be used to eliminate the $1/R$ term from the diffusion equation, leaving the expression

$$\frac{\partial C}{\partial T} = 2\frac{\partial^2 C}{\partial R^2} + \frac{\partial^2 C}{\partial Z^2} \tag{9.70}$$

which we apply *only* at $R = 0$. For the first column ($i = 0$) in the Z-sweep, the α, β and γ coefficients are as normal but the λ coefficients become

$$\lambda_1 = 0 \tag{9.71}$$

$$\lambda_2 = \frac{4}{h^2} - \frac{2}{\Delta T} \tag{9.72}$$

$$\lambda_3 = -\frac{4}{h^2} \tag{9.73}$$

Equation (9.70) may also be used as a boundary condition for the first point in every column in the R-sweep as an alternative to using $\frac{\partial C}{\partial R} = 0$.

9.2.4. *Flux*

The dimensionless current, J_i, at some point, i, on the electrode surface is equal to the concentration gradient normal to the surface at that point:

$$J_i = J_i^k = -\left(\frac{\partial C}{\partial Z}\right)_{Z=0} = -\left(\frac{C_{i,1}^{k-\frac{1}{2}} - C_{i,0}^{k-\frac{1}{2}}}{Z_1 - Z_0}\right) \tag{9.74}$$

Note that the current is calculated with the concentration values obtained just after the Z-sweep where the implicit diffusion is calculated perpendicular to the electrode surface.

The total dimensionless current of the entire disc, J, is found by integrating $J_i R$ across the electrode surface:

$$J = \int_0^1 J_i R \, dR \tag{9.75}$$

The integral of a continuous function, $f(x)$, may be approximated numerically using the trapezium rule [10]. The function is divided into a series of discrete intervals (discretised) and evaluated at each point, x. The area under the curve of a single interval is approximated as being trapezoidal, and the area of all such trapezoids is summed to give the integral. In general for a function discretised into N points,

$$\int_a^b f(x) \, dx \approx \sum_{i=1}^{N} \frac{f(x_i) + f(x_{i-1})}{2}(x_i - x_{i-1}) \tag{9.76}$$

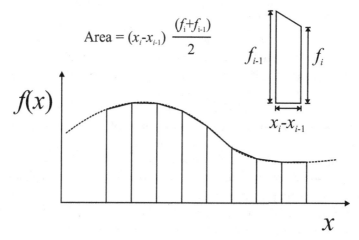

Fig. 9.6. The trapezium rule for numerical integration.

This process is illustrated in Figure 9.6. The accuracy depends on the interval size, Δx: the smaller this size, the more accurate the approximation [10].

In our case, the function J_i has already been evaluated at discrete intervals and so the total current is simply found from

$$J = \int_0^1 J_i R \, dR \approx \sum_{i=1}^{n_e-1} \frac{J_i R_i + J_{i-1} R_{i-1}}{2} (R_i - R_{i-1}) \qquad (9.77)$$

$$J \approx -\frac{1}{2} \sum_{i=1}^{n-1} \left[\frac{(C_{i,1}^{k-\frac{1}{2}} - C_{i,0}^{k-\frac{1}{2}}) R_i + (C_{i-1,1}^{k-\frac{1}{2}} + C_{i-1,0}^{k-\frac{1}{2}}) R_{i-1}}{Z_1 - Z_0} \right] (R_i - R_{i-1}) \qquad (9.78)$$

The real current at a particular timestep is then

$$I = 2\pi F D c_A^* r_e J \qquad (9.79)$$

9.3. Implementation

An implementation of a basic microdisc program for one-electron cyclic voltammetry written in C++ using the methods developed in this chapter is given in Appendix B.

9.3.1. *Multithreading*

In computing terms, a *process* is an instance of a computer program that is currently being executed. Each process has its own set of resources (such as memory) assigned to it by the operating system (OS) and the OS shields it from interactions from other processes running on the same machine. A process consists of one or more *threads* of execution which can each access the same shared pool of resources but not resources from outside the process. The OS gives the illusion of being able to perform many tasks simultaneously through the use of context switching; the OS specifies how much processor (CPU) time each thread receives per second depending on its priority, switching rapidly between them, giving the user the impression that they are all running at the same time. The important point here is that even though a computer can seem to perform many tasks simultaneously, a single processor core can only process a single thread at a time; however, a machine with multiple cores can process multiple threads simultaneously.

The nature of the ADI discretisation technique allows us to take advantage of this aspect of computer architecture to greatly increase the speed of simulation. Within a given sweep, each column or row may be evaluated completely independently as the results from one column/row in a given sweep do not in any way affect the results of another in the same sweep. Figure 9.7 illustrates the program flow using a multithreaded approach. A single master thread handles simulation setup (creating data structures, generating the grid spacings, etc.). This thread then spawns a number of other worker threads which simultaneously process the columns/rows during a sweep. The number of rows/columns that can be evaluated simultaneously is equal to the number of available processor cores; when a thread has finished processing a given row/column, it starts on the next unprocessed one until the sweep is complete.

Note that there is a certain amount of *computational overhead* associated with creating and using multiple threads, but this is more than compensated for by the increase in speed due to simultaneous processing.

At the time of writing, a high-end desktop machine can have 16-32 logical processor cores and so a roughly $16 - 32\times$ speed increase may be seen when using a multithreaded approach on such a machine compared to a single-threaded approach. A supercomputer may have an effectively unlimited number of cores so that all of the columns/rows in a sweep may be evaluated simultaneously; however, the execution speed is still limited by the speed at which a single column can be evaluated.

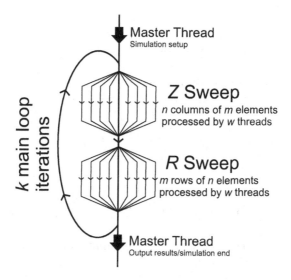

Fig. 9.7. Representation of the simulation in terms of threads.

The example program in Appendix B includes an implementation of multithreading using OpenMP.

9.3.2. *Testing Simulations: Specific tests*

As discussed for (hemi)spherical microelectrodes in Chapter 4, the response in cyclic voltammetry at microdiscs varies from a transient, peaked shape to a steady-state, sigmoidal one as the electrode radius and/or the scan rate are decreased, that is, as the dimensionless scan rate, $\sigma = Fr_e^2 v/(RTD)$, is decreased.

For large σ values (short $T_{\max}$ limit), the electrochemical response must coincide with that predicted for a macroelectrode so that the test criteria discussed in Chapter 4 apply.

In the transient regime with significant convergent diffusion (intermediate $T_{\max}$), the empirical Shoup and Szabo expression [11] provides accurate results (error smaller than 0.6%) for chronoamperometry under limiting current conditions

$$|J(T)| = \frac{2}{\pi}\left[0.7854 + \frac{0.8863}{2\sqrt{T}} + 0.2146\exp\left(-\frac{0.7823}{2\sqrt{T}}\right)\right] \qquad (9.80)$$

For cyclic voltammetry, the following empirical expression describes the value of the peak current of the forward peak for electrochemically reversible processes [12]

$$|J_p| = \frac{2}{\pi} \left(0.34 e^{-0.66\sqrt{\sigma}} + 0.66 - 0.13 e^{-11/\sqrt{\sigma}} + 0.351\sqrt{\sigma} \right) \tag{9.81}$$

Under steady state conditions (long T_{max} limit), the sigmoidal current-potential response of reversible one-electron transfers at ultramicrodiscs is given by

$$J_{ss} = -\frac{2}{\pi} \left(\frac{1 - \frac{c_B^*}{c_A^*} e^{\theta}}{1 + \frac{e^{\theta}}{d_B}} \right) \tag{9.82}$$

characterised by a plateau current of $|J_{disc,plateau}| = 2/\pi$ and a reversible half-wave potential of $\theta_{1/2,ss} = \ln(d_B)$ (as at (hemi)spherical ultramicroelectrodes).

For non-reversible transfers, a simple, approximate (yet accurate) expression is also available [13]

$$J_{ss} = -\frac{1}{2} \frac{K_{red} \left(1 - \frac{c_B^*}{c_A^*} e^{\theta} \right)}{1 + \frac{\pi}{4} K_{red} \left(1 + \frac{e^{\theta}}{d_B} \right)} \tag{9.83}$$

that in the limit of fully irreversible electrode reactions leads to the following expression for the half-wave potential

$$\theta_{1/2,irre} = \frac{\ln\left(\frac{\pi}{4} K^0\right)}{\alpha} \tag{9.84}$$

with $K^0 = k^0 r_e / D_A$, as defined in Chapter 4.

With regard to SWV at microdiscs, the values of the peak position and the half-peak width of fully-reversible electron transfers are the same as at the macroelectrodes (Subsection 3.5.2), since they do not depend on the size and shape of the electrode, and the peak current under steady state conditions is given by

$$|\Delta J_{p,ss}| = \frac{2}{\pi} \tanh\left(\frac{\theta_{SW}}{2}\right) \tag{9.85}$$

9.4. Microband Electrodes

A microband electrode is a conducting band that has a microscale width but a large, macroscale length, and is embedded in an insulating surface. In contrast to the microdisc, which we considered in terms of cylindrical

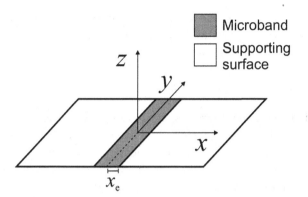

Fig. 9.8. The (x, y, z) Cartesian coordinate system used to model a microband electrode. The band half-width radius is x_e.

polar coordinates, the microband is most conveniently modelled in terms of three-dimensional Cartesian coordinates, as illustrated in Figure 9.8. Like the microdisc, this geometry also contains features that permit simplification to a two-dimensional model. As the length of the band is of macroscale, it is large compared to the thickness of the diffusion layer on the experimental time scale. We therefore make the approximation that the diffusional behaviour of the band does not vary across its length, i.e., that $\frac{\partial c}{\partial y} = 0$. Obviously in a real band, the behaviour at the band ends will differ from the behaviour at the middle of the band, but for a long band this leads to a negligible difference in the overall voltammetry.

In this way, the problem is reduced to two spatial dimensions, and we simply model a single (x, z)-plane [14–17]. Due to the line of mirror symmetry across the y-axis, we need only consider half of the band. The full three-dimensional current response is achieved by multiplying the two-dimensional response by $2l$ where l is the length of the band.

The spatial coordinates are normalised in terms of the half-width of the band, x_e:

$$X = \frac{x}{x_e} \tag{9.86}$$

$$Z = \frac{z}{x_e} \tag{9.87}$$

In two-dimensional Cartesian coordinates, the normalised version of Fick's second law is

$$\frac{\partial C}{\partial T} = \frac{\partial^2 C}{\partial X^2} + \frac{\partial^2 C}{\partial Z^2} \tag{9.88}$$

where $T = Dt/x_e^2$.

The simulation space and boundary conditions for a microband is exactly the same as that for a microdisc, as depicted in Figure 9.3, except for the change of coordinate system.

9.4.1. *Numerical solution*

For the microband, the discrete spatial grid used for the microdisc is still appropriate, and we also continue to solve using the ADI method. The mathematical techniques may be implemented in exactly the same way, the only change being different values for the coefficients α, β, γ and λ.

For the Z-implicit sweep, the terms in Eq. (9.88) are discretised as

$$\frac{\partial^2 C}{\partial Z^2} = \left(\frac{C_{i,j+1}^{k-\frac{1}{2}} - C_{i,j}^{k-\frac{1}{2}}}{\Delta Z_+} - \frac{C_{i,j}^{k-\frac{1}{2}} - C_{i,j-1}^{k-\frac{1}{2}}}{\Delta Z_-} \right) \frac{2}{\Delta Z_+ + \Delta Z_-} \qquad (9.89)$$

$$\frac{\partial^2 C}{\partial X^2} = \left(\frac{C_{i+1,j}^{k-1} - C_{i,j}^{k-1}}{\Delta X_+} - \frac{C_{i,j}^{k-1} - C_{i-1,j}^{k-1}}{\Delta X_-} \right) \frac{2}{\Delta X_+ + \Delta X_-} \qquad (9.90)$$

Therefore at each space point we have an equation of the form

$$\alpha C_{i,j-1}^{k-\frac{1}{2}} + \beta C_{i,j}^{k-\frac{1}{2}} + \gamma C_{i,j+1}^{k-\frac{1}{2}} = \lambda_1 C_{i-1,j}^{k-1} + \lambda_2 C_{i,j}^{k-1} + \lambda_3 C_{i+1,j}^{k-1} \qquad (9.91)$$

where

$$\alpha = \frac{2}{\Delta Z_+ + \Delta Z_-} \left(\frac{1}{\Delta Z_-} \right) \qquad (9.92)$$

$$\beta = \frac{2}{\Delta Z_+ + \Delta Z_-} \left(-\frac{1}{\Delta Z_+} - \frac{1}{\Delta Z_-} \right) - \frac{2}{\Delta T} \qquad (9.93)$$

$$\gamma = \frac{2}{\Delta Z_+ + \Delta Z_-} \left(\frac{1}{\Delta Z_+} \right) \qquad (9.94)$$

$$\lambda_1 = -\frac{2}{\Delta X_+ + \Delta X_-} \left(\frac{1}{\Delta X_-} \right) \qquad (9.95)$$

$$\lambda_2 = -\frac{2}{\Delta X_+ + \Delta X_-} \left(-\frac{1}{\Delta X_+} - \frac{1}{\Delta X_-} \right) - \frac{2}{\Delta T} \qquad (9.96)$$

$$\lambda_3 = -\frac{2}{\Delta X_+ + \Delta X_-} \left(\frac{1}{\Delta X_+} \right) \qquad (9.97)$$

For the X-sweep, at each space point we have an equation of the form

$$\alpha C^k_{i,j-1} + \beta C^k_{i,j} + \gamma C^k_{i,j+1} = \lambda_1 C^{k-\frac{1}{2}}_{i-1,j} + \lambda_2 C^{k-\frac{1}{2}}_{i,j} + \lambda_3 C^{k-\frac{1}{2}}_{i+1,j} \qquad (9.98)$$

where

$$\alpha = \frac{2}{\Delta X_+ + \Delta X_-} \left(\frac{1}{\Delta X_-} \right) \qquad (9.99)$$

$$\beta = \frac{2}{\Delta X_+ + \Delta X_-} \left(-\frac{1}{\Delta X_+} - \frac{1}{\Delta X_-} \right) - \frac{2}{\Delta T} \qquad (9.100)$$

$$\gamma = \frac{2}{\Delta X_+ + \Delta X_-} \left(\frac{1}{\Delta X_+} \right) \qquad (9.101)$$

$$\lambda_1 = -\frac{2}{\Delta Z_+ + \Delta Z_-} \left(\frac{1}{\Delta Z_-} \right) \qquad (9.102)$$

$$\lambda_2 = -\frac{2}{\Delta Z_+ + \Delta Z_-} \left(-\frac{1}{\Delta Z_+} - \frac{1}{\Delta Z_-} \right) - \frac{2}{\Delta T} \qquad (9.103)$$

$$\lambda_3 = -\frac{2}{\Delta Z_+ + \Delta Z_-} \left(\frac{1}{\Delta Z_+} \right) \qquad (9.104)$$

9.4.2. *Flux*

The current response of a microband electrode may be determined from:

$$I = 2Flj'_A \qquad (9.105)$$

where l is the length of the band in the y-direction and j' is the flux along a cross-section of the band in the (x, z)-plane. This assumes the surface is uniform for all planes parallel to the (x, z)-plane, which is valid as long as the band is long. The factor 2 arises because the simulation space considers only half of the band system. The value of j'_A is given by

$$j'_A = -D \int_{x=0}^{x=x_e} \frac{\partial c_A}{\partial z} \, dx \qquad (9.106)$$

Using the same procedure as we used for the microdisc in Subsection 9.1.3, we find that the dimensionless flux across the band cross-section is

$$J'_A = -\int_{X=0}^{X=1} \frac{\partial C_A}{\partial Z} \, dX \qquad (9.107)$$

Therefore the current is equal to

$$I = 2FlDc_A^* J_A'$$
(9.108)

where, by the trapezium rule,

$$J_A' = -\int_{X=0}^{X=1} \frac{\partial C_A}{\partial Z}\, dX$$

$$\approx -\frac{1}{2} \sum_{i=1}^{n_e-1} \left[\frac{(C_{i,1}^{k-\frac{1}{2}} - C_{i,0}^{k-\frac{1}{2}}) + (C_{i-1,1}^{k-\frac{1}{2}} - C_{i-1,0}^{k-\frac{1}{2}})}{Z_1 - Z_0} \right] (X_i - X_{i-1})$$
(9.109)

where the subscript i indicates the space step along the x-axis.

The numerical solution in diffusion-limited chronoamperometry can be checked by comparison with the expression derived by Szabo *et al.* [18]:

for $T < 1.6$

$$|J'| = \frac{1}{2} + \frac{1}{\sqrt{\pi T}}$$
(9.110)

for $T \geq 1.6$

$$|J'| = \frac{\sqrt{\pi}}{4} \frac{1}{\sqrt{T}} \exp\left(-0.2\sqrt{\pi T}\right) + \frac{\pi}{2} \frac{1}{\ln\left(5.2945 + 2.9972\sqrt{T}\right)}$$
(9.111)

References

[1] D. J. Gavaghan, An exponentially expanding mesh ideally suited to the fast and efficient simulation of diffusion processes at microdisc electrodes. 3. Application to voltammetry., *J. Electroanal. Chem.* **456**, 25–35, (1998).

[2] D. J. Gavaghan, An exponentially expanding mesh ideally suited to the fast and efficient simulation of diffusion processes at microdisc electrodes. 1. Derivation of the mesh., *J. Electroanal. Chem.* **456**, 1–12, (1998).

[3] D. F. Morton, K. W.; Mayers, *Numerical Solution of Partial Differential Equations: An Introduction*. (Cambridge University Press, Cambridge, 2005), 2nd edition.

[4] J. A. Alden and R. G. Compton, A general method for electrochemical simulations. I. Formulation of the strategy for two-dimensional simulations., *J. Phys. Chem. B.* **101**, 8941–8954, (1997).

[5] J. Heinze, Diffusion processes at finite (micro) disk electrodes solved by digital simulation., *J. Electroanal. Chem.* **124**, 73–86, (1981).

[6] J. W. Peaceman and H. H. Rachford, The numerical solution of parabolic and elliptical differential equations, *J. Soc. Ind. Appl. Math.* **3**, 28–41, (1955).

[7] D. J. Gavaghan and J. S. Rollett, Correction of boundary singularities in numerical simulation of time-dependent diffusion processes at unshielded disk electrodes., *J. Electroanal. Chem.* **295**, 1–14, (1990).

[8] D. Britz and J. Strutwolf, *Digital Simulation in Electrochemistry. 4th ed.* (Springer, Berlin, 2016).

[9] D. J. Gavaghan, How accurate is your two-dimensional numerical simulation? Part 1. An introduction., *J. Electroanal. Chem.* **420**, 147–158, (1997).

[10] W. H. Press, S. A. Teukolsky, W. T. Vetterling, and B. P. Flannery, *Numerical Recipes.* (Cambridge University Press, Cambridge, 2007), 3rd edition.

[11] D. Shoup and A. Szabo, Chronoamperometric current at finite disk electrodes, *J. Electroanal. Chem.* **140**, 237–245, (1982).

[12] R. G. Compton and C. E. Banks, *Understanding Voltammetry, 3rd ed.* (World Scientific Press, Singapore, UK, 2018).

[13] A. Molina, J. Gonzalez, E. O. Barnes, and R. G. Compton, Simple analytical equations for the current–potential curves at microelectrodes: a universal approach, *The Journal of Physical Chemistry C.* **118**(1), 346–356, (2013).

[14] J. A. Alden and R. G. Compton, A comparison of finite difference algorithms for the simulation of microband electrode problems with and without convective flow., *J. Electroanal. Chem.* **402**, 1–10, (1996).

[15] K. Aoki, Theory of ultramicroelectrodes., *Electroanal.* **5**, 627–39, (1993).

[16] R. G. Compton, A. C. Fisher, R. G. Wellington, P. J. Dobson, and P. A. Leigh, Hydrodynamic voltammetry with microelectrodes: channel microband electrodes; theory and experiment., *J. Phys. Chem.* **97**, 10410–15, (1993).

[17] I. Streeter, N. Fietkau, J. Del. Campo, R. Mas, F. X. Munoz, and R. G. Compton, Voltammetry at regular microband electrode arrays: Theory and experiment., *J. Phys. Chem. C.* **111**, 12058–12066, (2007).

[18] A. Szabo, D. K. Cope, D. E. Tallman, P. M. Kovach, and R. M. Wightman, Chronoamperometric current at hemicylinder and band microelectrodes: Theory and experiment, *J. Electroanal. Chem.* **217**, 417–423, (1987).

Chapter 10

Heterogeneous Surfaces

An electrochemically heterogeneous electrode is one where the electrochemical activity varies over the surface of the electrode. This broad classification encompasses a variety of electrode types [1, 2] including microelectrode arrays, partially blocked electrodes, electrodes made of composite materials, porous electrodes and electrodes modified with distributions of micro- and nanoscale electroactive particles. In this chapter, we extend the mathematical models developed in the previous chapter, in order to accurately simulate microelectrode arrays. Further, we explore the applications of a number of niche experimental systems, including partially blocked electrodes, highly ordered pyrolytic graphite, etc., and develop simulation models for them.

Microelectrode arrays are systems in which a number of micron-scale electrodes (usually of the same size and shape) are distributed over an inert supporting surface. When properly constructed, an array of microelectrodes wired in parallel offers the enhanced sensitivity observed of single microelectrodes but with the benefit of a higher total current output [3]. They find significant use in electrochemical analysis and sensor technology due to a number of advantages such as reduced ohmic drop, small capacitive charging currents and steady-state diffusion currents [4–6]. Parallel-wired microdisc arrays can also increase the electrochemical window in which one can perform experiments while retaining an easily measurable current output [7]. Arrays may also be wired such that each electrode is individually addressable [8, 9] which confers a high degree of spatial resolution. This allows the probing of signal transmission within a network of biological cells, and the possibility of sensing multiple analytes using different electrodes in the array [10].

A number of methods exist for fabricating microelectrode arrays [6] and a variety of array geometries are encountered with the most common being arrays of microdiscs and microbands. Microdiscs are most frequently arranged as a regularly distributed (i.e., a square or hexagonal lattice), flat array, although arrays of recessed discs are also used [11]. Microbands are typically employed as regular arrays wired in parallel [3], or as so-called interdigitated microband arrays, where two arrays of bands (each itself wired in parallel) are arranged in an alternating pattern. Interdigitated systems are often used in redox cyclic, generator-collector experiments [12]. Other, more exotic array types are occasionally encountered, including non-flat geometries, such as arrays of cylindrical microelectrodes [13].

A partially blocked electrode is a macroscale electrode that is partially covered in electrochemically inert microscale particles which block the diffusional path of electroactive solution-phase species to the electrode surface [14]. The inverse situation is an inert surface modified with a distribution of (usually hemispherical or spherical) electroactive nanoparticles; such systems are currently finding widespread use in electroanalysis [15].

Certain electrode materials display heterogeneous electrochemical behaviour with certain zones of the surface being more electroactive than other zones. This can arise, for example, in pyrolytic graphite which is uniform in composition but possesses two distinct crystallographic planes (edge and basal), both of which are electroactive but which each display different electrochemical rate constants [16].

10.1. Arrays of Microdisc Electrodes

In Chapter 9, we studied the problem of a single electroactive microdisc on an infinite supporting surface. Here we consider the situation where an array of such microdiscs are embedded in a surface in a regular distribution as illustrated in Figure 10.1. It is assumed that electroactivity only occurs at the microdisc electrodes, not on the supporting surface.

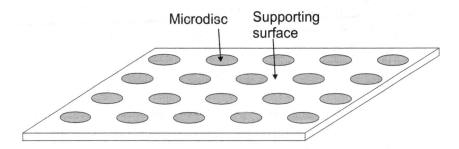

Fig. 10.1.   A regularly distributed array of microdisc electrodes.

We assume that the array is large, consisting of many microdiscs. Since the discs are regularly distributed, it is reasonable to suppose that on average each disc sits in an identical diffusional environment, with differences in environment of the outer edges of the array making a negligible contribution to the whole. We may therefore treat the surface as though it is composed of a large number of identical unit cells, each centred around a single disc. These cells may be arranged in, e.g., a hexagonal[1] array as shown in Figure 10.2. The centre-to-centre distance of two nearest neighbours is d, which is the same as the height of the hexagon (distance between two parallel sides). As each unit cell behaves identically, it is only necessary to simulate one of them; the current response from the whole surface is then simply the response of a single cell multiplied by the number of discs in the array. Even though simulating a single unit cell is a much smaller problem than simulating the whole surface, it is still a three-dimensional problem and is thus highly computationally demanding. When simulating an isolated microdisc, we relied on the rotational symmetry of the system to reduce the problem to a two-dimensional one. This symmetry is not present in our unit cell; however, a simplified solution is available through the use of the diffusion domain approximation.

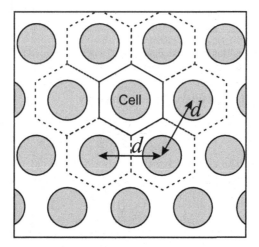

Fig. 10.2.   Hexagonal array of microdisc electrodes with centre-to-centre separation, d.

---

[1] Here we assume a hexagonal lattice, but this discussion applies equally to other regular distributions such as a square lattice.

10.1.1. *Diffusion domain approximation*

Under the diffusion domain approximation [1, 14, 17, 18], the hexagonal unit cell is approximated as a cylindrical cell of the same base area as shown in Figure 10.3.

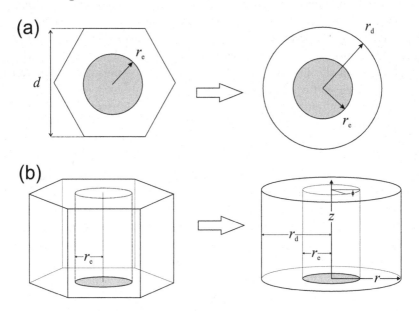

Fig. 10.3.   Transformation of a hexagonal cell to a cylindrical cell of the same base area: (a) top-down view; (b) three-dimensional unit cell view.

The area, A, of the hexagonal base is given by

$$A = \frac{\sqrt{3}}{2}d^2 \tag{10.1}$$

The radius, r_d, of the circular base may therefore be calculated by equating the areas of the two shapes:

$$\pi r_d^2 = \frac{\sqrt{3}}{2}d^2 \tag{10.2}$$

so

$$r_d \approx 0.525d \tag{10.3}$$

Previous studies have demonstrated that the results of simulations using the diffusion domain approximation show very good agreement with experiments [19].

10.1.2. *Implementation*

As with the isolated microdisc simulations in Chapter 9, we here consider the simulation of the cyclic voltammetry of a simple fully reversible one-electron reduction. For an array, since each unit cell is identical, the concentrations of the electroactive species will necessarily be the same on either side of the cell boundary and there can be no flux of electroactive material across the boundary. After using the diffusion domain approximation, this boundary is at a distance $r = r_d$, therefore

$$\left(\frac{\partial c}{\partial r}\right)_{r=r_d} = 0 \qquad (10.4)$$

Apart from this change in boundary condition, the simulation space for a microdisc array is exactly the same as that used for the single microdisc, depicted in Figure 9.3.

The cell radius is normalised against the disc radius:

$$R_d = \frac{r_d}{r_e} \qquad (10.5)$$

where R_d is the normalised cell radius. In order to perform an array simulation, this dimensionless distance must be determined. While it can be explicitly specified by the simulation user, it is usually more appropriate to calculate it based on some other aspect of the array geometry. A commonly used method is to allow the user to specify a surface coverage, Θ, which is defined generally for any partially active electrode surface as the ratio of electroactive area to total area. For an array of discs this may be written as

$$\Theta = \frac{N\pi r_e^2}{A} \qquad (10.6)$$

where N is the number of discs in the array, and A is the total substrate area. The coverage may equivalently be determined from

$$\Theta = \frac{\pi r_e^2}{\pi r_d^2} \qquad (10.7)$$

The normalised radius, R_d, may therefore be calculated from a specified value of surface coverage by

$$R_d = \sqrt{1/\Theta} \qquad (10.8)$$

10.1.2.1. *Simulation procedure*

The simulation procedure for a microdisc array is almost identical to that developed using the ADI method for a single microdisc in Chapter 9; there are only two significant changes. First, as we have discussed, is the imposition of the diffusion domain boundary at $r = r_d$ (where before we had a bulk concentration boundary condition at $r = r_{max}$). In discrete form, this condition is represented as $C^k_{n-1,j} - C^k_{n-2,j} = 0$ and this is implemented by using the following coefficients for the last point of every row, $i = n - 1$, in the R-sweep:

$$\alpha_{n-1} = -1, \qquad \beta_{n-1} = 1, \qquad \gamma_{n-1} = 0, \qquad \delta_{n-1} = 0 \qquad (10.9)$$

Additionally, it is necessary to alter the grid spacing in the R-direction to allow for a greater density of spatial points in the region of this outer boundary which may be achieved using an expanding-contracting patching scheme as described in Subsection 4.1.1, such that there are the greatest density of spatial points at $R = 0$, $R = 1$ and $R = R_d$. The increments, h_i ($= R_{i+1} - R_i$), are given by

$$0 \leq R \leq \frac{1}{2}, \qquad h_i = h_0 \omega^i \qquad (10.10)$$

$$\frac{1}{2} < R \leq 1, \qquad h_i = h_0 \omega^{n_e - i} \qquad (10.11)$$

$$1 < R \leq \frac{R_d + 1}{2}, \qquad h_i = h_0 \omega^{i - n_e} \qquad (10.12)$$

$$\frac{R_d + 1}{2} < R \leq R_d, \qquad h_i = h_0 \omega^{n - i} \qquad (10.13)$$

The grid spacing for the Z-direction may remain as it was for a single microdisc, i.e., a high density at the electrode surface, expanding up to the outer boundary, as the outer boundary in that direction is still a bulk concentration boundary at semi-infinite distance.

10.1.3. *Diffusional modes*

It is useful at this point to examine the diffusional and voltammetric behaviour of a microdisc array as it varies with the surface coverage (or equivalently with the disc-to-disc separation) and disc size. It is observed that there are four limiting cases of behaviour; this classification applies generally to all partially active electrodes and was originally based on observations of partially blocked electrode systems [1, 2].

The category that any given experiment/simulation will fall into depends largely upon the distance that the electroactive species diffuse on the time scale of the experiment ($\sim \sqrt{Dt}$) relative to the average separation between neighbouring discs, as well as upon the average disc radius. The time scale may be controlled in, for example, cyclic voltammetry, through the scan rate (high scan rate leads to short experimental time and low scan rate leads to a long experimental time); or in potential step chronoamperometry, by the duration of the potential pulse applied.

The diffusional behaviour observed in each case is illustrated schematically in Figure 10.4, accompanied by an example of the simulated linear sweep voltammetry that can be expected under each regime. The example voltammograms are simulated based on a system with a surface coverage $\Theta = 0.01$, disc radius $r_e = 1$ μm and diffusion coefficient $D_A = 10^{-5}$ cm^2 s^{-1}, and vary only in terms of the voltammetric scan rate, ν.

Case 1 behaviour, depicted in Figure 10.4(a), describes diffusion that is perpendicular to the surface of the electroactive site. This behaviour is observed when the size of the discs is large compared with $\sqrt{Dt}$. This can correspond to either a large (macroscale) disc, or to an experiment conducted over an extremely short time scale, such as a cyclic voltammetry with a very fast scan rate; the faster the scan rate, the smaller the discs can be while still observing Case 1 behaviour. In the limit of fully reversible kinetics, the voltammetric response reflects linear diffusion to the electroactive area such that the peak current, I_p, is given by the Randles–Ševčík equation:

$$I_p = -0.4463 \, \mathrm{F} A_{ea} c^* \sqrt{\frac{\mathrm{F} D \nu}{\mathcal{R} T}} \qquad (10.14)$$

where $A_{ea} = N \pi r_e^2$ is the total electroactive surface area of the microdisc array.

Case 2 behaviour, depicted in Figure 10.4(b), is observed when the size of the discs is small compared with $\sqrt{Dt}$. Under this category, each disc behaves as a microelectrode and diffusion to it is convergent rather than linear. In both Cases 1 and 2, neighbouring discs are sufficiently far apart that they are effectively diffusionally independent in the experimental time scale. The voltammetric response in Case 2 is therefore that of a single isolated microdisc electrode, scaled by the number of discs in the array;

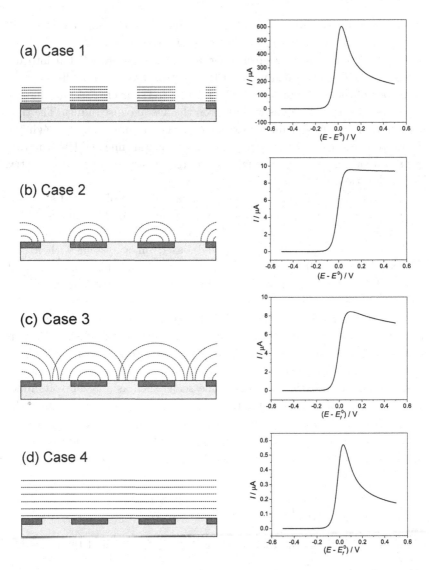

Fig. 10.4. The four limiting cases of diffusional behaviour to an array of microdisc electrodes.

steady-state voltammetry is therefore observed for such arrays (assuming the scan rate is not too high). It should be emphasised again that diffusional independence in this case depends not on the absolute size of the inter-disc separation, d, but upon the ratio of this distance to the distance that electroactive species can diffuse on the time scale of the experiment ($\sim \sqrt{Dt}$); specifically, diffusional independence is achieved in Cases 1 and 2 if $d > \sqrt{Dt}$. Consequently, the concept of diffusional independence can only apply to an electrode array for a finite amount of time [20].

Case 3 behaviour, depicted in Figure 10.4(c), is also observed when the size of the discs is small and when the inter-disc separation is less than that in Case 2 (again relative to $\sqrt{Dt}$). In this category, diffusion to each individual disc is still convergent, but the diffusion fields of neighbouring discs begin to overlap somewhat, a situation that is sometimes referred to as shielding. The current density is therefore lower than in Case 2 where each disc is effectively independent. The voltammetric response is intermediate in character between the sigmoidal, steady-state voltammetry observed for Case 2 and the peaked voltammetry observed in Case 1 (and Case 4). When designing microelectrode arrays for use as, e.g., certain types of sensors, it is often desirable to maintain diffusional independence and thus avoid Case 3 (shielding) behaviour in order to achieve a linear signal enhancement without additionally complicating the response [20].

Case 4 behaviour, depicted in Figure 10.4(d), represents the extreme limit of Case 3, where the spacing between adjacent discs is very small, such that there is very strong overlap of neighbouring diffusion fields and $d < \sqrt{Dt}$. In this case, diffusion is linear to the entire surface and the voltammetric response shows a well-defined peak. Perhaps surprisingly, in this limit, the peak current is given by the Randles–Ševčík equation where A is the total geometric area of the electrode system (active and inert); it is the same as it would be with the entire surface electroactive![2]

Although the peak current in Case 4 does not vary with the surface coverage of electroactive material, Θ is seen to affect the apparent electrochemical rate constant, k_{app}. It has been shown [2, 17] that so long as the diffusional behaviour is strictly in the Case 4 limit

$$k_{app}^0 = k^0 \Theta \tag{10.15}$$

such that a lower surface coverage results in slower apparent kinetics. In the limit of fully reversible electrode kinetics, this has no influence on the

---

[2] This classification scheme may be extended to a fifth case [3]. In Case 5, the experimental time scale is very long (very slow scan rate) so that diffusion to the entire substrate is convergent, giving sigmoidal voltammetry. This behaviour is most readily observed experimentally when the supporting substrate itself is of microscale and the electroactive particles are of sub-micron size.

observed voltammetry; however, this does not hold for the irreversible limit. In the latter case, the peak potential, E_p, is given by

$$E_p = E_f^0 - \frac{RT}{\alpha F}\left[0.780 - \ln(\Theta k^0) + \ln\left(\sqrt{\frac{\alpha F D\nu}{RT}}\right)\right] \qquad (10.16)$$

Equivalently, in normalised units this is

$$\theta_p = \frac{1}{\alpha}\left[\ln\left(\frac{\Theta K^0}{\sqrt{\sigma\alpha}}\right) - 0.780\right] \qquad (10.17)$$

A theoretical study [21] has demonstrated that this relationship also applies to non-flat electrode geometries, such as inert surfaces modified with electroactive nanoparticles, except that in that case, the apparent rate constant is given by $k_{app}^0 = k^0\Psi$ where Ψ is the ratio of electroactive surface area to geometric surface area of the electrode. Whereas necessarily $\Theta \leq 1$, Ψ is not so restricted, and may in fact have a value greater than 1. In that case, the apparent rate constant is seen to increase compared to a fully active macroelectrode, and the peak position, E_p, shifts towards E_f^0, giving the *appearance* of a catalytic effect. Again this assumes that the diffusional behaviour is in Case 4.

10.1.4. *Random arrays*

We now turn our attention to *randomly* distributed arrays of microdisc electrodes as illustrated in Figure 10.5; though they are not as commonly encountered as regularly distributed microdisc arrays, techniques do exist for their fabrication [22] and so here we consider the simulation of such arrays. Though the specific example of a randomly distributed *microdisc* array is of limited utility, the techniques for generating a random distribution of particles are applicable to a range of electrochemical problems.

Fig. 10.5.   A randomly distributed array of microdisc electrodes.

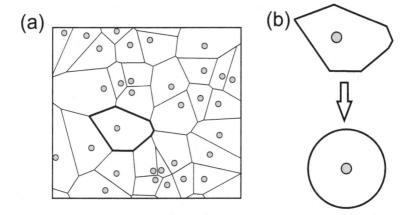

Fig. 10.6. (a) Surface with randomly distributed microdisc electrodes divided into Voronoi cells; (b) diffusion domain approximation used to transform a Voronoi cell into a cylindrical cell of equivalent base area.

A randomised algorithm could be used that would take as input the disc radius, r_e, the total area of the array system, A, and the surface coverage, Θ, and would generate a surface with randomly distributed microdiscs; alternatively the number of particles, N, could be specified in place of r_e since

$$N = \frac{A\Theta}{\pi r_e^2} \tag{10.18}$$

Figure 10.6(a) illustrates a section of a surface generated by such an algorithm. In the figure, each disc[3] sits in its own cell; the cell's boundary encompasses the set of points that is closer to that particular disc than to any other disc, and is called a Voronoi cell (the cell boundaries may also be considered as bisections of the distance between the nearest neighbours). The surface is therefore composed of N domains of varying size, each of which is an irregular polygon of area A_n. The sides of the polygon are situated midway between two adjacent discs, thus to a good approximation, they can be considered to be no-flux boundaries.

In the same manner as for the regularly distributed array, we can use the diffusion domain approximation to transform each Voronoi cell into a cylindrical cell of the same base area, A_n, and of radius $r_{d,n}$, as illustrated

---

[3] We assume that each disc is of the same radius; the simulation of randomly distributed arrays of particles of different sizes is beyond the scope of this text.

in Figure 10.6(b), thus reducing the problem of simulating each cell from three dimensions to two dimensions. As each cell is now effectively of a different size, it would seem that in order to calculate the current response, I, for the whole surface, it is necessary to first generate the surface with the randomised algorithm, calculate the area of each individual cell, simulate each one and finally sum their respective current responses, I_n:

$$I = \sum_{n=1}^{N} I_n \tag{10.19}$$

Using this approach it would take approximately N times longer to simulate a single randomly distributed array than it would to simulate a regular array, which is clearly impractical, as N is likely to be very large.

Fortunately, a set of equations are available for the statistical treatment of nearest-neighbour distance in a random distribution of particles [18, 23]. For a particular value of r_e, knowledge of N and A allows us to determine the distribution of diffusion domain sizes (in terms of domain radius, r_d) present on the electrode surface. The probability, $P(r_d)$, of finding domains of radius r_d is given by the equation (a Poisson distribution):[4]

$$P(r_d) = \frac{2\pi N r_d}{A} \exp\left(\frac{-\pi N r_d^2}{A}\right) \tag{10.20}$$

This distribution is illustrated graphically in Figure 10.7.

The expectation value (mean) domain radius, $\langle r_d \rangle$, is given by [18]

$$\langle r_d \rangle = \sqrt{\frac{A}{4N}} \tag{10.21}$$

To find the total current response of the surface, I, we need to solve the integral

$$I = N \int_0^\infty I(r_d) P(r_d) \, dr_d \tag{10.22}$$

where $I(r_d)$ is the current response of a domain of radius r_d. The probability of domains with radii greater than $3\langle r_d \rangle$ is so small that their current contribution is negligible [18], and the upper limit of infinity can thus be

---

[4] It should be noted that this allows for the case of overlapping discs. However the approximation remains valid as long as the degree overlap is not excessive (as long as the coverage is not too high).

Fig. 10.7. The probability function given by Eq. (10.20) for a range of values of r_e where $\Theta = 0.1$.

replaced by $3\langle r_d \rangle$. This integral can be approximated numerically using the trapezium rule, which for equal intervals leads to:

$$I = \sum_{m=1}^{M} \frac{1}{2} \left[I(r_{d,m} + \Delta r_d) P(r_{d,m} + \Delta r_d) + I(r_{d,m}) P(r_{d,m}) \right] \Delta r_d \quad (10.23)$$

where Δr_d is the interval size and M is the total number of intervals between 0 and $3\langle r_d \rangle$ such that $M = 3\langle r_d \rangle / \Delta r_d$. The smaller the value of Δr_d, the more closely this will approximate the true value of the integral.

In normalised units, Eq. (10.20) is

$$P(R_d) = \frac{2\pi N R_d}{A'} \exp\left(\frac{-\pi N R_d^2}{A'} \right) \quad (10.24)$$

where R_d is the normalised domain radius and A' is the normalised electrode area given by

$$A' = \frac{A}{r_e^2} \quad (10.25)$$

We may similarly transform Eq. (10.23):

$$J = \sum_{m=1}^{M} \frac{1}{2} \left[J(R_{d,m} + \Delta R_d) P(R_{d,m} + \Delta R_d) + J(R_{d,m}) P(R_{d,m}) \right] \Delta R_d$$

(10.26)

where now

$$M = \frac{3 \langle R_d \rangle}{\Delta R_d} = \frac{3}{\Delta R_d} \sqrt{\frac{A'}{4N}}$$

(10.27)

To summarise then, simulation of a surface with a random distribution of microdiscs with any given set of values of Θ and A' (or N and A') consists of the following:

(1) Select an interval size, ΔR_d, and from this calculate the number of intervals, M.
(2) For each value of R_d thus generated (ΔR_d, $2\Delta R_d$, ..., $M\Delta R_d$), simulate a domain of that radius and record the normalised current response, $J(R_d)$.
(3) Use these values of $J(R_d)$, and the values of $P(R_d)$ given by Eq. (10.24) to solve for J in Eq. (10.26).

In this way, the current response (voltammetry) of each domain size is weighted appropriately according to its probability of occurring on the surface.

10.1.4.1. *Partially blocked electrodes*

A partially blocked electrode (PBE) is a macroscale electrode that is partially covered in particles of some inert blocking material. Electrochemistry at PBEs presents an intriguing problem as the blocking modifies the rates of diffusive and kinetic flux to the electrode surface. Not only can these changes lead to inaccurate electrochemical interpretation, but in some circumstances, the blocked electrode can be mistaken for an unblocked one with the wrong mechanism [24].

If we approximate each of the inert blocking particles as being disc-shaped and of the same size, the modelling of a PBE is only very slightly different from modelling a random array of microelectrodes. In the latter case we considered an array of electroactive discs on an inert surface whereas for a PBE we consider an array of inert discs on an electroactive surface. The simple solution then is to use exactly the same simulation model as for the random array of microdiscs except that the surface boundary conditions

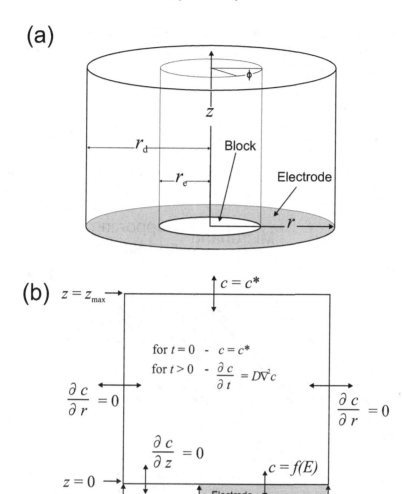

Fig. 10.8. (a) Unit cell for the partially blocked electrode model; (b) simulation space for the PBE.

are inversed; the boundary at $(z = 0, r \leq r_e)$ is now set to use a zero-flux boundary condition, and the boundary at $(z = 0, r > r_e)$ is now set to use a potential-dependent boundary condition. Figure 10.8 illustrates the unit cell and the simulation space for a PBE.

If the size of the blocking particles is relatively small such that the diffusion can be described as being in Case 4, the peak current of voltammetry of the blocked surface will be that given by the Randles–Ševčík equation, Eq. (10.14) and in the limit of irreversible kinetics, the peak potential will be that given by Eq. (10.16).

10.2. Microband Arrays

In a microband electrode array, a series of microband electrodes lie across a surface in a parallel arrangement [3] as illustrated in Figure 10.9. It is assumed that electroactivity is confined to the electrodes; the supporting surface is inert in this respect.

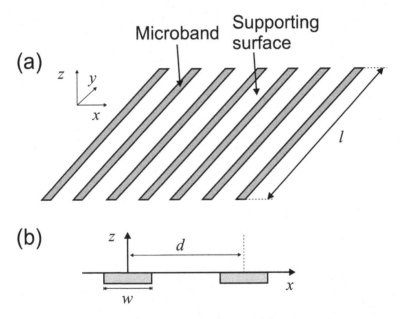

Fig. 10.9. (a) Schematic diagram of a regularly distributed array of microband electrodes. (b) Cross-section through the (x, z)-plane.

For the purposes of modelling this array, we assume that each band is identical, with the same length, l, and width, w, and with the same centre-to-centre spacing, d, between each adjacent band. Consequently, each band sits in the centre of an identical unit cell, as shown in Figure 10.10. This assumes that there are many bands in the array; bands at the outer edge of the array will obviously experience a different environment than those at the centre; however, if there are a large number of bands, the effect of this

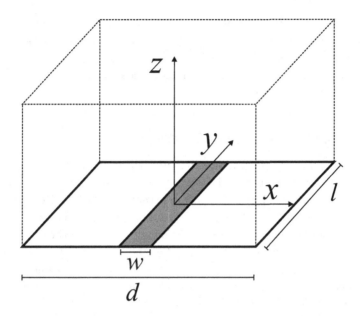

Fig. 10.10. The unit cell for a single microband electrode in an array.

difference on the observed voltammetry will be negligible. If we assume that each band experiences an identical environment, and therefore that each unit cell is identical, it is only necessary to simulate a single unit cell. The total voltammetric response of the entire array is found by scaling the current recorded at the single band by the number of bands in the array. In contrast to the microdisc array, the microband unit cell already contains the requisite symmetry necessary to reduce the problem to two dimensions, so no transformation akin to the diffusion domain approximation is necessary.

The boundary between two adjacent cells, which is at a distance $x_d = d/2$ from the centre of each electrode, is actually a line of mirror symmetry. As each side of the boundary is identical, there can necessarily be no flux across it and so the boundary condition at $x = x_d$ is

$$\left(\frac{\partial c}{\partial x}\right)_{x=x_d} = 0 \tag{10.28}$$

Apart from this change in the boundary condition, the simulation space for a microband array is exactly the same as for the single microband, depicted in Figure 9.8. As with the microdisc array, it is necessary to alter the spatial grid in the X-direction so that there is a high concentration of spatial points at this outer boundary to ensure accurate simulation.

Note that the classification of diffusional modes developed in Subsection 10.1.3 also applies to arrays of microbands [3] and so is useful when analysing their behaviour.

10.2.1. *Highly ordered pyrolytic graphite*

Highly ordered pyrolytic graphite (HOPG) and its derivatives are popular electrode materials for use in electroanalysis and are examples of materials that have electrochemical rate constants that vary across their surfaces. HOPG consists of stacked planes of graphite with an interlayer spacing of 3.35 Å, and possesses two electrochemically distinct surfaces: the basal plane which lies parallel to the graphite plane, and the edge plane which lies perpendicular to it [16, 25]. The edge plane surface displays an electrochemical rate constant (k_0^{edge}) that is several orders of magnitude faster than the basal plane (k_0^{basal}) for most redox couples. As a consequence, electrode surfaces of widely varying character can be prepared from a sample of HOPG by cleaving it either parallel to or perpendicular to the plane of the graphite sheets. In the former case the electrode will show mostly basal plane character (slow, irreversible kinetics), whereas in the latter case it will have mostly edge plane character (fast, reversible kinetics). No cut will be perfect, however, and therefore the surface will always take the form of a stepped structure as shown schematically in Figure 10.11(a), with long terraces of the dominant plane and small steps (impurities) of the minor

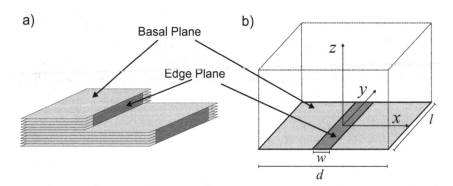

Fig. 10.11.   (a) Schematic of the layered structure of HOPG; (b) flat unit cell used to model BPPG.

plane. Note that some recent work suggests that so-called pristine HOPG shows basal plane zones with fast electrode kinetics but that this is rapidly passivated soon after preparation.

The difference in the electrochemical rate constants is such that in the case of a surface with mostly edge-plane character, called edge plane pyrolytic graphite (EPPG), the small regions of basal plane are effectively inert. These regions act as thin band-shaped blocks, in the same manner as the blocking discs in the PBE model (Subsubsection 10.1.4.1), which as we saw have a relatively small effect on the voltammetry, causing a very small shift in the peak position, according to Eq. (10.16), but no change to the peak current. In the case of basal plane pyrolytic graphite (BPPG); however, the situation is more interesting: the relative size of the edge plane defects has a significant effect on the peak potential of the voltammetry.[5] Therefore it is informative to simulate the BPPG system to better understand this effect.

As the BPPG surface consists of wide flat terraces of basal plane separated by narrow steps of edge plane, it is reasonable to treat the surface as though it is completely flat [26]. This simplification allows us to simulate voltammetry of a BPPG surface using the model that we have already developed for microband arrays, with only a small change to the surface boundary conditions. The unit cell for the BPPG model is shown in Figure 10.11(b). Under the microband array model, the surface boundary conditions are given by

$$\text{for } z = 0, \ x \leq x_e: \qquad c = f(k_0, E) \qquad (10.29)$$

$$\text{for } z = 0, \ x > x_e: \qquad \frac{\partial c}{\partial z} = 0 \qquad (10.30)$$

where $f(k_0, E)$ is a kinetic and potential-dependent boundary condition with a form defined by, for example, the Butler–Volmer or Marcus–Hush kinetic models (see Chapter 4). The new surface boundary conditions for the BPPG model are

$$\text{for } z = 0, \ x \leq x_e: \qquad c = f(k_0^{\text{edge}}, E) \qquad (10.31)$$

$$\text{for } z = 0, \ x > x_e: \qquad c = f(k_0^{\text{basal}}, E) \qquad (10.32)$$

---

[5] In the extreme limit of this model, corresponding to very thin edge bands and wide basal bands, voltammetry with two distinct peaks in the forward sweep may even be observed [26].

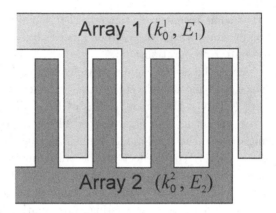

Fig. 10.12.   Schematic diagram of an interdigitated microband system.

10.2.2. *Interdigitated microbands*

Interdigitated microband arrays (IDAs) are another example of an electrode system with spatially variant electrode kinetics. The system consists of two separate arrays of microband electrodes, each wired in parallel, arranged in an alternating pattern, as illustrated in Figure 10.12. The two arrays may be made of different electroactive materials, allowing for different kinetic properties, and the potential applied at each may be independently regulated. The primary application of IDAs is as generator-collector systems in which one of the arrays (the generator) is used to oxidise or reduce some target analyte, which is then transported to the second array (the collector) where it undergoes further oxidation/reduction:

$$\text{Generator:} \quad \text{A} \pm \text{e}^- \rightleftharpoons \text{B} \qquad\qquad (10.33)$$

$$\text{Collector:} \quad \text{B} \pm \text{e}^- \rightleftharpoons \text{C} \qquad\qquad (10.34)$$

In the case where $\text{C} = \text{A}$, redox cycling occurs, where electroactive species rapidly cycle between the generator and collector electrodes [27], greatly enhancing the measured current and therefore the sensitivity. Further, it is possible to use such a setup to simultaneously measure both the concentration and diffusion coefficient of a target analyte [28].

It is possible to develop a simulation model for the IDA system by recognising that, like some other systems under study in this chapter, it has translational symmetry. The simulation space for the IDA is a unit cell that encompasses half of each of two neighbouring bands and the space in

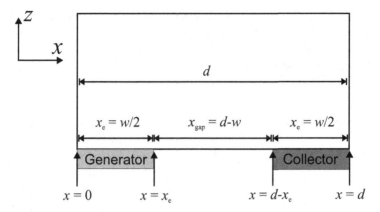

Fig. 10.13. Simulation space for an interdigitated microband system.

between them, as shown in Figure 10.13. Note that here for simplicity, we assume that both kinds of band have the same width, w, though this need not be the case.

The simulation model is similar to that used for the microband array but with altered surface ($z = 0$) boundary conditions and an x-grid spacing modified to take account of the newly introduced singularity at $x = d - x_e$. The surface boundary conditions are different for each species and vary with x according to the following (where it is assumed that A is electro-reduced to B at the generator and B is further reduced to C at the collector)

$$0 \leq x \leq x_e \quad \begin{cases} D_A \left(\frac{\partial c_A}{\partial z}\right)_{z=0} = k_{red}^{(i)} c_A(i,0) - k_{ox}^{(i)} c_B(i,0) \\ D_B \left(\frac{\partial c_B}{\partial z}\right)_{z=0} = -D_A \left(\frac{\partial c_A}{\partial z}\right)_{z=0} \\ D_C \left(\frac{\partial c_Y}{\partial z}\right)_{z=0} = 0 \end{cases} \quad (10.35)$$

$$x_e < x \leq d - x_e \quad \begin{cases} D_A \left(\frac{\partial c_A}{\partial z}\right)_{z=0} = 0 \\ D_B \left(\frac{\partial c_B}{\partial z}\right)_{z=0} = 0 \\ D_C \left(\frac{\partial c_C}{\partial z}\right)_{z=0} = 0 \end{cases} \quad (10.36)$$

$$d - x_e < x \leq d \quad \begin{cases} D_A \left(\frac{\partial c_A}{\partial z}\right)_{z=0} = 0 \\ D_B \left(\frac{\partial c_B}{\partial z}\right)_{z=0} = k_{red}^{(ii)} c_B(i,0) - k_{ox}^{(ii)} c_C(i,0) \\ D_C \left(\frac{\partial c_C}{\partial z}\right)_{z=0} = -D_B \left(\frac{\partial c_B}{\partial z}\right)_{z=0} \end{cases} \quad (10.37)$$

where $k_{red/ox}^{(i)/(ii)}$ are the rate constants for the reduction and oxidation at the generator, (i), and collector, (ii).

The spatial grid in the x-direction should use an expanding-contracting patching scheme such that the point density is highest at the two edges of each electrode (at $x = 0$, $x = x_e$, $x = d - x_e$ and $x = d$).

The current is calculated from

$$I = 2Fl(j'_A + j'_B) \qquad (10.38)$$

where the flux is integrated separately across the surface of each band:

$$j'_A = -D_A \int_{x=0}^{x=x_e} \left(\frac{\partial c_A}{\partial z} \right)_{z=0} dx \qquad (10.39)$$

$$j'_B = -D_B \int_{x=d-x_e}^{x=d} \left(\frac{\partial c_B}{\partial z} \right)_{z=0} dx \qquad (10.40)$$

10.3. Porous Electrodes

We consider an electrode system that consists of a macroelectrode covered with an insulating film that contains (non-interconnected) cylindrical micropores such that electroactive species in solution can only access the conducting electrode surface by diffusion through the pores [29]. This system is illustrated schematically in Figure 10.14(a). For the purposes of modelling, we assume that the pores are of uniform radius, r_e, and depth, z_e, and that there are a large number of them regularly distributed across the surface. Further, we assume that the percentage of the surface occupied by the pores is given by the surface coverage, $\Theta = N\pi r_e^2/A$, where N is the total number of pores and A is the geometric area of the macroelectrode. This allows for the application of the diffusion domain approximation (Subsection 10.1.1) under which we can consider each pore to sit at the centre of an identical cylindrical unit cell, with a radius r_d which is calculated from:

$$r_d = \frac{r_e}{\sqrt{\Theta}} \qquad (10.41)$$

The unit cell and coordinates are illustrated in Figure 10.14(b). As with the array of microdiscs model, the unit cell is cylindrically symmetrical about an axis that passes through the centre of the pore, perpendicular to the electrode surface. The problem may thus be reduced from a three-dimensional one to a two-dimensional one. As with the microdisc electrode, this is a two-dimensional cylindrical polar coordinate system, and Fick's second law in this space is given by Eq. (9.6). The simulation space for the

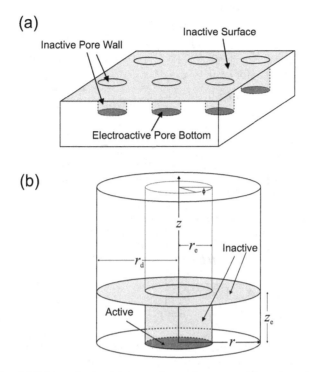

Fig. 10.14. (a) Schematic of an electroactive surface covered in an inactive porous film; (b) cylindrical unit cell of a single pore derived from the diffusion domain approximation.

unit cell with its attendant boundary conditions is shown in Figure 10.15. The complete set of boundary conditions is

$$
\begin{aligned}
0 \leq z \leq z_{\max}, \ r = 0: && \left(\frac{\partial c}{\partial r}\right)_{r=0} &= 0 \\
0 \leq z \leq z_{e}, \ r = r_{e}: && \left(\frac{\partial c}{\partial r}\right)_{r=r_{e}} &= 0 \\
z_{e} < z \leq z_{\max}, \ r = r_{d}: && \left(\frac{\partial c}{\partial r}\right)_{r=r_{d}} &= 0 \\
z = 0, \ 0 \leq r \leq r_{e}: && c &= f(E) \\
z = z_{e}, \ r_{e} \leq r \leq r_{d}: && \left(\frac{\partial c}{\partial z}\right)_{z=z_{e}} &= 0 \\
z = z_{\max}, \ 0 \leq r \leq r_{d}: && c &= c^{*}
\end{aligned}
\tag{10.42}
$$

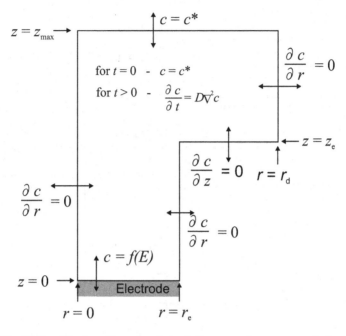

Fig. 10.15. The two-dimensional simulation space for a porous electrode.

This model differs from any previous two-dimensional model that we have studied in that the simulation space is not rectangular; the zone defined by $(r_e < r \leq r_d,\ 0 \leq z < z_e)$ is outside the bounds of the space. In terms of computational implementation, the container that stores the concentration grid may still be rectangular; one simply sets the initial concentration of all species at all points in this exclusion zone to zero. In addition, any coefficients for the Thomas algorithm $(\alpha_{i,j}, \beta_{i,j}, \gamma_{i,j})$ that refer to spatial points inside this zone are also set to zero, and the discretised boundary conditions derived from (10.42) are applied in the appropriate places. The current is calculated in exactly the same manner as for a microdisc electrode.

The voltammetric behaviour of this system is controlled by the size of the diffusion layer $(\sim \sqrt{Dt})$ relative to the pore depth, z_e, and radius, r_e [29]. If the diffusion layer thickness is less than the pore depth, diffusion to the electrode surface is necessarily planar as the layer is confined to the interior of the pores and there is no possibility of radial diffusion, regardless of the size of r_e. As a consequence, conventionally macroelectrode voltammetry is seen even if the radius of the pore is of micron-scale. If the size of the diffusion layer is greater than the depth of the pore then the voltammetric behaviour becomes sensitive to both r_e and r_d.

10.4. Conclusions

This chapter has demonstrated that, in spite of being computationally more demanding, a wide variety of electrode geometries can be successfully modelled; many complex three-dimensional systems can be simulated in a reasonable amount of time and with acceptable accuracy by carefully considering any symmetry inherent in the system and exploiting this to reduce the problem to a more manageable number of dimensions. From there, all that is frequently needed is an appropriate description of the boundary conditions of the system and a careful consideration of the discrete spatial gridding to ensure a high density of points in the region of any finite spatial boundary or singularity.

References

[1] R. G. Compton and C. E. Banks, *Understanding Voltammetry, 3rd ed.* (World Scientific Press, Singapore, UK, 2018).
[2] T. J. Davies, C. E. Banks, and R. G. Compton, Voltammetry at spatially heterogeneous electrodes., *J. Solid State Electrochem.* **9**, 797–808, (2005).
[3] I. Streeter, N. Fietkau, J. Del. Campo, R. Mas, F. X. Munoz, and R. G. Compton, Voltammetry at regular microband electrode arrays: Theory and experiment., *J. Phys. Chem. C.* **111**, 12058–12066, (2007).
[4] H. P. Wu, Fabrication and characterization of a new class of microelectrode arrays exhibiting steady-state current behavior., *Anal. Chem.* **65**, 1643–1646, (1993).
[5] J. Magee, L. Joseph and J. Osteryoung, Fabrication and characterization of glassy carbon linear array electrodes., *Anal. Chem.* **61**, 2124–2126, (1989).
[6] X.-J. Huang, A. M. O'Mahony, and R. G. Compton, Microelectrode arrays for electrochemistry: Approaches to fabrication., *Small.* **5**, 776–788, (2009).
[7] F. G. Chevallier and R. G. Compton, Widening the voltammetric window using regular arrays of microdisk electrodes., *Electroanal.* **19**, 1741–1745, (2007).
[8] G. Grancharov, E. Khosravi, D. Wood, A. Turton, and R. Kataky, Individually addressable recessed gold microelectrode arrays with monolayers of thio-cyclodextrin nanocavities., *Analyst.* **130**, 1351–1357, (2005).
[9] B. Zhang, K. L. Adams, S. J. Luber, D. J. Eves, M. L. Heien, and A. G. Ewing, Spatially and temporally resolved single-cell exocytosis utilizing individually addressable carbon microelectrode arrays., *Anal. Chem.* **80**, 1394–1400, (2008).

[10] M. P. Nagale and I. Fritsch, Individually addressable, submicrometer band electrode arrays. 1. Fabrication from multilayered materials., *Anal. Chem.* **70**, 2902–2907, (1998).

[11] F. A. Aguiar, A. J. Gallant, M. C. Rosamond, A. Rhodes, D. Wood, and R. Kataky, Conical recessed gold microelectrode arrays produced during photolithographic methods: Characterisation and causes., *Electrochem. Commun.* **9**, 879–885, (2007).

[12] E. O. Barnes, G. E. M. Lewis, S. E. C. Dale, F. Marken, and R. G. Compton, Generator-collector double electrode systems: A review., *Analyst.* **137**, 1068–1081, (2012).

[13] H. Xu, K. Malladi, C. Wang, L. Kulinsky, M. Song, and M. Madou, Carbon post-microarrays for glucose sensors., *Biosens. Bioelectron.* **23**(11), 1637–1644, (2008).

[14] B. A. Brookes, T. J. Davies, A. C. Fisher, R. G. Evans, S. J. Wilkins, K. Yunus, J. D. Wadhawan, and R. G. Compton, Computational and experimental study of the cyclic voltammetry response of partially blocked electrodes. Part 1. Nonoverlapping, uniformly distributed blocking systems., *J. Phys. Chem. B.* **107**, 1616–1627, (2003).

[15] F. W. Campbell and R. G. Compton, The use of nanoparticles in electroanalysis: An updated review., *Anal. Bioanal. Chem.* **396**, 241–259, (2010).

[16] R. L. McCreery, Advanced carbon electrode materials for molecular electrochemistry., *Chem. Rev.* **108**, 2646–2687, (2008).

[17] C. Amatore, J. M. Savéant, and D. Tessier, Charge transfer at partially blocked surfaces. A model for the case of microscopic active and inactive sites., *J. Electroanal. Chem.* **147**, 39–51, (1983).

[18] T. J. Davies, B. A. Brookes, A. C. Fisher, K. Yunus, S. J. Wilkins, P. R. Greene, J. D. Wadhawan, and R. G. Compton, A computational and experimental study of the cyclic voltammetry response of partially blocked electrodes. Part II: Randomly distributed and overlapping blocking systems., *J. Phys. Chem. B.* **107**, 6431–6444, (2003).

[19] T. J. Davies, S. Ward-Jones, C. E. Banks, J. Del. Campo, R. Mas, F. X. Munoz, and R. G. Compton, The cyclic and linear sweep voltammetry of regular arrays of microdisc electrodes: Fitting of experimental data., *J. Electroanal. Chem.* **585**, 51–62, (2005).

[20] D. Menshykau, X. Huang, N. Rees, F. del Campo, F. Munoz, and R. G. Compton, Investigating the concept of diffusional independence. Potential step transients at nano- and micro-electrode arrays: Theory and experiment., *Analyst.* **134**, 343–348, (2009).

[21] K. R. Ward, M. Gara, N. S. Lawrence, R. S. Hartshorne, and R. G. Compton, Nanoparticle modified electrodes can show an apparent increase in electrode kinetics due solely to altered surface geometry: The effective electrochemical rate constant for non-flat and non-uniform electrode surfaces., *J. Electroanal. Chem.* **695**, 1–9, (2013).

[22] S. Fletcher and M. D. Horne, Random assemblies of microelectrodes (RAM electrodes) for electrochemical studies., *Electrochem. Commun.* **1**, 502–512, (1999).

[23] S. Chandrasekhar, Stochastic problems in physics and astronomy, *Rev. Mod. Phys.* **15**, 1–89, (1943).

[24] O. Contamin and E. Levart, Characterization and identification of partially blocked electrodes., *J. Electroanal. Chem.* **136**, 259–70, (1982).

[25] H. Chang and A. J. Bard, Observation and characterization by scanning tunneling microscopy of structures generated by cleaving highly oriented pyrolytic graphite., *Langmuir.* **7**, 1143–53, (1991).

[26] K. R. Ward, N. S. Lawrence, R. S. Hartshorne, and R. G. Compton, The theory of cyclic voltammetry of electrochemically heterogeneous surfaces: Comparison of different models for surface geometry and applications to highly ordered pyrolytic graphite., *Phys. Chem. Chem. Phys.* **14**, 7264–7275, (2012).

[27] O. Niwa, M. Morita, and H. Tabei, Electrochemical behavior of reversible redox species at interdigitated array electrodes with different geometries: Consideration of redox cycling and collection efficiency., *Anal. Chem.* **62**, 447–52, (1990).

[28] K. Aoki and M. Tanaka, Time-dependence of diffusion-controlled currents of a soluble redox couple at interdigitated microarray electrodes., *J. Electroanal. Chem.* **266**, 11–20, (1989).

[29] F. G. Chevallier, L. Jiang, T. G. J. Jones, and R. G. Compton, Mathematical modelling and numerical simulation of cyclic voltammetry at an electrode covered with an insulating film containing cylindrical micropores., *J. Electroanal. Chem.* **587**, 254–262, (2006).

Chapter 11

Stochastic Electrochemistry

In the previous chapters, we considered electrochemical systems in which mass transport can be described by partial differential equations. In particular, we discussed the diffusion equation and used it to model mass transport in the absence of migration and convection, and the Nernst-Planck- and Nernst-Planck-Poisson equations which both extend diffusion equation to include migration induced by electrostatic effects. All three equations yield concentration profiles $c(\vec{r}, t)$ that describe the distribution of an analyte across the solution as a density in terms of the amount n per volume V:

$$c(x, y, z, t) = c(\vec{r}, t) = \frac{\partial n(\vec{r}, t)}{\partial V} \tag{11.1}$$

which is typically expressed either in units of molar concentration M or mol m^{-3}. Since n is evaluated for an infinitesimally small volume V, n is a real number in $\mathbb{R}^{\geq 0}$ and does not account for the discrete nature of the analyte which, of course, is comprised of individual molecules that can be counted. Mathematically, the forms of the three partial differential equations (diffusion-, Nernst-Planck, and Nernst-Planck-Poisson equation), being first order in time and second order in space, demand the concentration to be continuous in time and space and continuously differentiable in space, which a distribution of individual molecules is of course not. The requirements imposed on analyte distribution by the structure of the mentioned three mass transport equations however hardly limit their applicability to describe most experimental systems: If we, for instance, consider a typical current of 10 nA measured at a micro-disk electrode, we find that the total number of electrons transferred per second is as large as:

$$n_e = \frac{10 \text{ nA}}{e_0} \approx 6.24 \cdot 10^{10} \ s^{-1} \tag{11.2}$$

Even if sampled at a relatively high rate of 1 kHz, i.e. 1000 measurements per second, more than 60 million electron transfers are recorded per measurement. Following the law of large numbers, the description of the mass transport via the concentration (11.1) is hence an excellent approximation despite the neglect of the discrete nature of charge and the analyte. So why is there any interest in stochastic mass transport in electrochemistry?

The last decades have seen significant advances in the capabilities of the potentiostats available to electrochemists: Modern potentiostats allow the measurement of currents at picoampere resolution while thousands of individual measurements can be recorded per second. Great advances have been made in *sensitivity* and *bandwidth*! Such advanced experimental set-ups are now widely used in academia and industry, and can be acquired as complete systems from commercial suppliers or can be custom-built [1] from individual components to meet more specific experimental requirements.

Access to high-bandwidth and low-current potentiostats has opened new opportunities in electrochemistry which had previously been obstructed by a lack of suitable measurement instruments. If smaller electrodes, faster measurement rates, or lower analyte concentrations are used, continuous models become invalid as the law of large numbers breaks down and fluctuations no longer average out. Instead, the randomness that is ubiquitous at the molecular scale is reflected in Faradaic current measurements: Discrete charge transfers, Brownian motion, ad- and desorption, thermal noise, and many other stochastic processes leave their fingerprint in measured currents [2]. Let us for instance consider a current of 100 fA that is recorded at a nano-electrode. In this case, the electrons transferred per second amount to:

$$n_e = \frac{100 \text{ fA}}{e_0} \approx 624,000 \ s^{-1} \qquad (11.3)$$

If sampled at a typical rate of 1 kHz, each sample reflects an average of only 624 electrons which is inevitably affected by the stochastic nature of the reaction process. The ability to resolve and analyse these fluctuations has sparked entirely new developments in electrochemistry which are, most notably, the nano-impact technique[1] and electrochemical correlation spectroscopy [4].

In this chapter, we approach the topic of stochastic electrochemistry from two different perspectives: First, we explore, in a stochastic context,

---

[1] This technique is sometimes referred to as single-entity electrochemistry or collision electrochemistry. A broad introduction to the topic can be found in [3].

what conclusions can be drawn from the partial differential equations we have discussed so far in this book. We revisit prior simulation approaches and show how these methods can be used in the different context of a stochastic analysis. Second, we introduce an entirely different modelling approach, the modelling via random walks: Theory is detailed and set in context related to partial differential equations.

11.1. Stochastic Electrochemistry and Partial Differential Equations

Consider a typical experimental set-up in stochastic electrochemistry, a nano-impact experiment. An electrode is set in contact with a solution comprising an electrolyte and a suspension of a nanomaterial which is present in the form of individual nanoparticles that move freely in the solution by virtue of their Brownian motion. Individual particles interact with the electrode in a way that if they reach the region in which electron tunnelling is likely, particles react while individual reactions are observed as 'spikes' in the electrode current. These particle reactions may for instance include an irreversible oxidation or reduction of the particle, or its electro-dissolution. Such *destructive* nano-impact measurements are at the heart of current research in stochastic electrochemistry as they have evolved to a versatile tool in electrochemistry. Applications are broad and include the characterisation of populations of silver nanoparticles [5], which are electro-dissolved or at least fully oxidised upon the impact at the electrode while the transferred charge per individual particle impact corresponds to the particle volume, and the detection influenza viruses [6], which are tagged with silver nanoparticles to allow visibility.

Once the data has been recorded in a nano-impact experiment, it can be analysed to extract various quantities of interest with the particle concentration often being one of the first values to be determined. The simulations we discussed in the previous chapters of this book have yielded electrode currents or fluxes of analyte at the electrode surface from which, for instance via equations like the Randles-Ševčík equation, we could infer the analyte concentration. The data obtained from a nano-impact experiment carried out to determine a concentration is however a stochastic sequence of points in time at which current spikes occurred and must be analysed or simulated in a different fashion.

A common approach to conduct such analysis employs the same partial differential equations we solved earlier in this book. However, the concentration is herein reinterpreted in terms of low concentrations of analyte molecules: Rather than the definition (11.1) where we interpret as the derivative of amount n [mol] with respect to volume V [m^3], we now consider a probability density p [m^{-3}] of finding a single analyte entity in a certain volume. This probability density is normalised with respect to the entire solution, that is if we integrate p over all space we must yield a probability $P = 1$ of finding the analyte entity. We hence define:

$$p(\vec{r}, t) = \frac{\partial P}{\partial V} \quad \text{with} \quad \int d\vec{r}\, p(\vec{r}, t) = 1 \tag{11.4}$$

where P is the probability of finding the analyte entity in a certain finite volume and where we demand that p is continuous in time and space and continuously differentiable in space. We will later see that the latter requirement regarding the continuity of p and $\partial_t p$ is useful when we determine solutions for p.

The probability density p can be used akin to probability densities we know from stochastics. If, for instance, we want to determine the average position $\bar{x}$ of an analyte along the x axis and p is known, we can evaluate the expression:

$$\bar{x}(t) = \int d\vec{r}\, x\, p(\vec{r}, t) \tag{11.5}$$

where we weight $p(x, y, z)$ with the position in x. All other moments $\overline{x^n}$, $\overline{y^n}$, and $\overline{z^n}$ can be determined analogously. Similar to the calculations we know from stochastics, we can calculate the probability of finding a *particular* analyte within a volume V_1 and write:

$$P_{V_1}(t) = \int_{V_1} d\vec{r}\, p(\vec{r}, t) \tag{11.6}$$

In a real experiment there is of course more than just one analyte entity in solution and generalisations are needed to use the above concept of a probability density to model an electrochemical system. To this end, we extend our definition of the probability density to model an ensemble of identical analyte entities by simple multiplication with the total number N_{tot} of analyte entities in the entire system. The average number of analyte entities that are located in V_1 is hence:

$$\overline{N}_{V_1}(t) = N_{tot} \int_{V_1} d\vec{r}\, p(\vec{r}, t)$$

$$= N_{tot} P_{V_1}(t) \tag{11.7}$$

Having revisited the definition of concentration $c(\vec{r}, t)$ in the limit of low concentrations and having established a stochastic equivalent in terms of the probability density $p(\vec{r}, t)$, we can now turn to solving mass transport equations in an electrochemical system. To this end, we consider the diffusion equation[2] in the light of our definitions of $c(\vec{r}, t)$ and $p(\vec{r}, t)$:

$$\frac{\partial c(\vec{r}, t)}{\partial t} = D\Delta c(\vec{r}, t) \tag{11.8}$$

As we have discussed previously, diffusion equation models the mass transport of the analyte by virtue of its Brownian motion and does not include any interactions between the diffusing analytes. This can be seen in the structure of the equation: Different solutions to diffusion equation can be added to each other while their sum remains a solution to the equation. In terms of our probability densities, this implies that probability densities calculated for different analyte entities can be summed up. To solve diffusion equation at low concentrations we hence solely have to replace $c(\vec{r}, t)$ with $p(\vec{r}, t)$, which includes Nernst- and Butler-Volmer type boundary conditions. If, however, we instead consider the electrostatic Nernst-Planck equation:

$$\frac{\partial c(\vec{r}, t)}{\partial t} = D\Delta c(\vec{r}, t) + \frac{DzF}{\mathcal{RT}}\nabla(c(\vec{r}, t)\nabla\phi) \tag{11.9}$$

we find that such superposition of solutions cannot be done in the same fashion as the positions of *all* charged analytes determine the motion of *each* analyte through their contributions to the electrostatic potential ϕ. Solutions for such systems of interacting particles hence require additional simplifying assumptions or different modelling approaches such as the random walk approach which we will discuss later in this chapter.

11.1.1. *Example: Chronoamperometry at Low Concentrations*

In this section, we come back to the above nano-impact experiment and consider the problem of a suspension of electroactive nanoparticles impinging on an electrode where they are electro-dissolved. To establish a solution

---

[2] In our example, we choose the diffusion coefficient D to be a constant. While this is a good approximation for small nanoparticles, for other systems it may be required to include the effect of near-wall hindered diffusion. The diffusion coefficient of the analyte is then a function of space as the Brownian motion is hydro-dynamically retarded in the immediate proximity of solid boundaries. In terms of the interpretation of $c(\vec{r}, t)$ as $p(\vec{r}, t)$, this does however not affect our modelling approach.

in one dimension, we consider the following system: Nano-particles are dispersed in a solution, diffuse freely, and upon contact with a macro-electrode, particles are consumed while a charge $\overline{q_p}$ is transferred to the electrode. Individual reactions are then recorded as a spike in the electrode current after a suitable potential is applied to the electrode at a given time.

At the beginning of the experiment and prior to the application of the potential, the solution comprises a concentration c^* of nano-particles which are homogeneously dispersed on a time average. The latter is the key assumption that enables us to define the initial state of the system and hence the boundary conditions subject to which we solve the diffusion equation. At the beginning of the experiment the particle concentration is not affected by the presence of the electrode and due to the stochastic nature of the particle mass transport, particles are randomly distributed. If we divide all space in the solution in equally-sized volumes, there will be volumes in which there are more particles and others in which there are less particles. The nano-particle concentration at any specific point in time is hence never homogeneous. The time average of the particle concentration is however constant throughout the entire solution. In an experiment, there are no means to experimentally determine the positions of all the nanoparticles at the beginning of the experiment to use their distribution in a simulation. Given the large number of particles, the time average is a very good approximation and we hence set:

$$\overline{p}(\vec{r}, t \leq 0) = p(\vec{r}, t = 0) = \text{const.} \qquad (11.10)$$

On the time scale of a typical experiment, the reaction at the electrode has no impact on the distribution of particles in the bulk solution and we limit our analysis to the volume V in the immediate proximity of the electrode surface. As in Eq. (2.20), we set the boundaries of this volume in a distance d from the electrode surface:

$$d = 6\sqrt{Dt_{max}} \qquad (11.11)$$

where t_{max} is the duration of the experiment following the application of the potential. Using the volume of the simulated space, we normalise the probability densities of the nanoparticles and define its initial value $p(\vec{r}, t = 0)$:

$$p(\vec{r}, t = 0) = \frac{1}{V} \quad \text{and} \quad N_{tot} = c^* N_A V \qquad (11.12)$$

where c^* is given in units of mol m^{-3}.

Having defined the initial concentration of the nanoparticles in solution, we have found all boundary conditions that we require to solve this mass transport problem. We first write the one-dimensional linear diffusion equation in terms of the probability density:

$$\frac{\partial p(x,t)}{\partial t} = D \frac{\partial^2 p(x,t)}{\partial x^2} \tag{11.13}$$

Since particles react immediately, entirely, and irreversibly upon reaching the electrode surface, the probability of finding a particle at the electrode surface must vanish at any time and hence:

$$p(x=0,t) = 0 \tag{11.14}$$

which in combination with Eq. (11.12) provides a complete set of boundary conditions. In terms of its mathematical structure, this problem of a single partial differential equation and two boundary conditions is identical to the problem of the chronoamperometry of a molecular analyte at high concentrations we have discussed in Chapter 4, and can be solved analogously. There are, however, differences in the interpretation of the flux at the electrode. Whereas in Chapter 4 we calculated an electrical current on the basis of the concentration c^*, we are here interested in the rate of particles reacting at the electrode as a function of the probability density p and the number of particles N_{tot} that are initially present in the simulated volume. To this end, we define a flux density p_f of the probability density p akin to Fick's first law:

$$p_f(t) = -D \frac{\partial p(x,t)}{\partial x}\bigg|_{x=0} \tag{11.15}$$

This expression features units Hz m^{-2} and provides the probability of an individual particle to react at the electrode per time and electrode area. We can now calculate the probability of a particle reacting at the electrode between the beginning of the experiment, i.e. the moment in which the electrode is set from a floating to the working potential, and the time t by integrating p_f with respect to time:

$$P_f(t) = \int_0^t d\tau \, p_f(\tau) \tag{11.16}$$

where τ denotes the variable of time, to which respect we integrate, to avoid confusion with the upper integration boundary t. The average cumulative number of particles that react at the electrode until time t can then be calculated by simply multiplying $P_f(t)$ with the initial number of nanoparticles present in solution at the beginning of the experiment:

$$\overline{N}(t) = N_{tot} A P_f(t)$$

$$= N_{tot} A \int_0^t d\tau \, p_f(\tau)$$

$$= -N_{tot} AD \int_0^t d\tau \frac{\partial p(x, \tau)}{\partial x}\bigg|_{x=0} \quad (11.17)$$

where A denotes the surface area of the macro-electrode. For a Cottrell-type system like the one we discussed in Chapter 4, we can substitute the Cottrell equation to obtain:

$$\overline{N}(t) = N_{tot} A p^* \sqrt{\frac{D}{\pi}} \int_0^t d\tau \frac{1}{\sqrt{\tau}}$$

$$= N_{tot} x_{max} \sqrt{\frac{D}{\pi}} \int_0^t d\tau \frac{1}{\sqrt{\tau}}$$

$$= 2 N_{tot} x_{max} \sqrt{\frac{Dt}{\pi}} \quad (11.18)$$

where we have used:

$$p(\text{all } x, t = 0) = p^* = \frac{1}{V} = \frac{1}{A x_{max}} \quad (11.19)$$

Figure 11.1 presents schematic experimental and simulated data to illustrate the comparison between experiment and simulation. The upper part of the figure shows the current $I(t)$ as a function of time where we see spikes, each of which corresponds to an individual particle impacting at the electrode, while the amount of charge transferred during each of these impacts averages to $\overline{q_p}$. This experimental data can straightforwardly be transformed into a graph of the cumulated number of impacts. We integrate the current with respect to time and subsequently divide by $\overline{q_p}$:

$$N^{exp}(t) = \frac{1}{\overline{q_p}} \int_0^t d\tau \, I(\tau) \quad (11.20)$$

which yields the staircase graph shown in the lower part of the figure. The dashed line depicts a simulation result obtained by the here-discussed approach via partial differential equations which, in contrast to the experimental results, does not feature stochastically-occurring steps but the

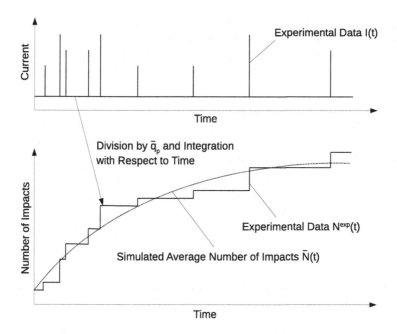

Fig. 11.1. Schematic illustration of the electrode current in a typical destructive nano-impact experiment (top). Normalisation of the experimental data by the average charge $\overline{q_p}$ transferred per impact event and integration with respect to time yield the staircase graph (bottom) of the cumulated number of particle impacts. The dashed line depicts $\overline{N}(t)$ as obtained from the simulation. Please note that fictitious data is shown to illustrate the comparison between theory and experiment. The different sizes of spikes reflects the different nanoparticle sizes present in solution.

average number of impacts. For a better comparison between experimental and the simulated average data, the experiment can hence be repeated multiple times and the obtained $I(t)$ can be averaged. The staircase plot of the averaged measured data then approaches the simulated data asymptotically with each repetition of the experiment.

The comparison of experimental and simulated data in Figure 11.1 illustrates the strengths and weaknesses of the simulation approach via partial differential equations. On the one hand, *average* concentrations in the form of probability densities and *average* reaction rates are an easily-accessible and practical way to rationalise experimental data and to extract essential physical quantities such as the particle concentration. On the other hand, the simulations of averaged data hides the stochastic nature of the reaction process: For systems in which an analysis can be based on underlying stochastic processes reflected in fluctuations in the electrode current, no conclusions can be drawn directly from the simulation. In the following part of this chapter, we will therefore turn to a different simulation approach, the random walk method, which will enable us to simulate these stochastic properties and model stochastic processes in much greater detail.

11.1.1.1. *Implementation in C++*

The partial differential equations and boundary conditions we solve for this problem feature the structure of the equations we solved in Chapter 3, where we established a computational model for diffusion equation in one dimension subject to a Nernstian boundary condition. The implementation is hence achieved along the lines of the example code given in that chapter. For better accuracy, the code can be advanced using expanding space and time grids as introduced in Chapter 4.

11.2. Stochastic Electrochemistry and Stochastic Differential Equations

Next, we approach stochastic electrochemistry from a different and generic angle, and attack the problem of theoretical modelling by means of stochastic differential equations. Previous chapters and the above section have described electrochemical systems through macroscopic mass transport models such as Fick's first- and second laws. Both describe the temporal evolution of the distribution of the analyte in the limit of an innumerably large number of analyte entities while relying on the calculation of gradients in this distribution. To this end, we have expressed distributions either as concentrations or as probability densities after the generalisation in the context of low concentrations where the latter fulfil the same requirements in terms of the calculation of gradients. These macroscopic equations allowed us to model average currents and analyte distributions which is generally sufficient for all experiments if either large electrode currents (see Eq. (11.2)) are expected or the knowledge of the average current in a stochastic system is sufficient. If, however, the character of the fluctuations seen in the electrode current or the pathways of individual analyte and product entities (molecules, nano-particles, enzymes, viruses, ...) are required, more detailed microscopic modelling approaches must be pursued.

Figure 11.2 shows a scheme of a modelling approach at the single-entity level. We herein first establish a stochastic model for the mass transport of the *individual* analyte and its interactions with the environment which

Fig. 11.2.   Scheme of the approach to stochastic simulations at the single-analyte level.

can, for instance, be a reaction upon contact with an electrode, migration in an electric field, or an electrostatic attraction or repulsion in the presence of other charged analytes. Simulations are then run while key features of each analyte like its oxidation state and its position are sampled and the behaviour of the entire macroscopic system is inferred. The following subsections will first provide a theoretical background of random walk simulations in the context of the diffusion equation. We will next turn to the general definition of a random walk, and finally we discuss the example of reversible voltammetry at a macroelectrode and its implementation in C++.

11.2.1.   *Theoretical Background*

This section introduces the theoretical background of random walk simulations in the context of diffusion equations. We discuss how the Brownian motion of an individual analyte can be modelled stochastically and illustrate the validity of the random walk approximation via a comparison with the macroscopic diffusion equation. In our discussion, we use various concepts from theoretical physics, in particular Langevin dynamics and the equipartition theorem, to provide a general understanding of the physics of the random walk. We give a broad overview but do not cover the underlying mathematics exhaustively. The interested reader is referred to standard textbooks in theoretical physics for further details.

We divide our introduction into subsections, each of which introduces a different aspect of the theoretical background before we finally combine the gained knowledge to derive a stochastic model for the Brownian motion.

11.2.1.1. *Friction*

To model the Brownian motion of a molecule, we must understand the friction it experiences during its motion through the solution. This friction arises from interactions with surrounding solvent molecules that are displaced and rearranged while the molecule travels through the solution and is typically defined via a one-dimensional Newtonian equation of motion. We write out the equation of motion of a molecule that is solely affected by an external Force and the friction with the solvent:

$$m\frac{d^2x(t)}{dt^2} = F - \gamma\frac{dx(t)}{dt} \tag{11.21}$$

where F is an external force which is constant in time and space and γ is the friction coefficient we define through this equation. Since the external force is set constant in time and space, the acceleration term on the left side of this equation converges to zero as the molecule approaches a constant speed over time. The equation then simplifies to:

$$0 = F - \gamma\frac{dx(t)}{dt} \tag{11.22}$$

which we can integrate to:

$$x(t) = \frac{F}{\gamma}t + \text{const.} \tag{11.23}$$

The molecule hence travels at a constant velocity of:

$$v(t) = \frac{dx(t)}{dt} = \frac{F}{\gamma} \tag{11.24}$$

which, possibly counterintuitively, only depends on the strength of the external force and the friction coefficient but not on the molecules' mass as gravity has been excluded as a force.

For a perfect sphere in a viscous solution, the friction coefficient can be derived from the Navier-Stokes equation in the limit of a small Reynold's number. The flow is then laminar as it is typically seen at the micro- and nano-scale. This solution is usually referred to Stokes' Law:

$$\gamma = 6\pi r \eta \qquad (11.25)$$

where r is the radius of the sphere and η is the dynamic viscosity of the solution. Stokes' law is very commonly used to describe the friction of a molecule in a solvent despite the molecule not being a perfect sphere. In this case r is called the Stokes radius.

Below in this section, we will come back to the topic of friction and use the definition of the friction coefficient in Eq. (11.21) and Stokes law in further derivations.

11.2.1.2. *The Equipartition Theorem*

The equipartition theorem is one of the key elements we require to develop a mathematical model for Brownian motion. To this end, let us revisit introductory quantum mechanics and consider a canonical ensemble with the Hamiltonian:

$$H(x) = H(x_1, x_2, \ldots, x_f) \qquad (11.26)$$

where the x_i are f degrees of freedom and coordinates in phase space, that is, generalised positions and generalised momenta. For such systems, the equipartition theorem states that:

$$\overline{x_i \frac{\partial H(x)}{\partial x_j}} = \delta_{ij} k_B \mathcal{T} \qquad (11.27)$$

if:

$$x_i \, exp\left(-\frac{H(x)}{k_B \mathcal{T}}\right) = 0 \quad \text{for} \quad x_i \longrightarrow \pm\infty \qquad (11.28)$$

where the overline represents the expected value and δ_{ij} is the Kronecker delta. δ_{ij} equals unity if $i = j$ and zero otherwise.

To derive equations on which we later build our model, we consider the example of a molecule that can move freely in one dimension and features the Hamiltonian:

$$H(p) = \frac{p^2}{2m} \qquad (11.29)$$

where p is the momentum of the molecule in one dimension. The equipartition theorem then yields:

$$\overline{p\frac{\partial H(p)}{\partial p}} = \frac{1}{m}\overline{p^2} = k_B T \qquad (11.30)$$

for this particular degree of freedom: the molecule's motion in x. Substitution of $p = mv$ and division by two then gives an average value for the kinetic energy of the molecule as a function of temperature:

$$\overline{E_{kin}} = \frac{1}{2}m\overline{v^2} = \frac{k_B T}{2} \qquad (11.31)$$

or:

$$\overline{v^2} = \frac{k_B T}{m} \qquad (11.32)$$

Eq. (11.32) is an important result we require below.

11.2.1.3. The Langevin Equation

The following section turns to the Langevin equation, a stochastic differential equation that is central to the description of any Brownian motion at the molecular scale. It describes the temporal evolution of a subset of relatively 'slowly' evolving degrees of freedom of a system while the other relatively 'fast' changing degrees of freedom are accounted for by the stochastic nature of the equation. In the case of Brownian motion in a solution, these slowly evolving degrees of freedom are the coordinates of the position of the molecule whose diffusion is modelled in x, y, and z, and the fast changing degrees of freedom are the numerous collisions of the molecule with solvent molecules which happen on a much faster time scale than the (in this context) slow motion of the molecule through the solution. In one dimension, the Langevin Equation can be written as:

$$m\frac{d^2 x(t)}{dt^2} = F(x(t), t) - \gamma\frac{dx(t)}{dt} + f(t) \qquad (11.33)$$

where $F(x,t)$ is an external force which, for instance, can result from the electric field of the electrode and which henceforth we set to zero,[3] and

---

[3] The absence of external forces is an excellent approximation for most electrochemical systems. Since experiments are typically conducted at electrolyte concentrations high enough to screen the electric field of the electrode in solution, the remaining electric fields that arise from other charges average out. Furthermore, the gravitational force is significantly smaller than the forces that result from the energy available to the analyte in terms of the equipartition theorem for its movement in parallel to the gravitational force.

$f(t)$ is a fluctuating force that can be modelled as a stochastic process. A closer look at the equation reveals that it bears significant similarity to a Newtonian equation of motion: On the left hand side we have a term that accounts for the acceleration of the molecule while the terms on the right represent the different forces the molecule experiences. We further recognise the term $-\gamma \frac{dx(t)}{dt}$ from the above paragraph on friction, which accounts for the friction force and points in the opposite direction than all other forces the molecule experiences. The force $f(t)$ is a stochastic force that accounts for the combined effect of all fast changing degrees of freedom, i.e. all collisions with solvent molecules. This force $f(t)$ features various important characteristics. First, $f(t)$ averages out over time:

$$\overline{f(t)} = 0 \tag{11.34}$$

as the molecule collides with solvent molecules on all sides and the fluctuating force therefore features no preferred direction. Second, the forces that result from the individual collisions with solvent molecules are entirely uncorrelated:

$$\overline{f(t)f(t')} = \Gamma\delta(t - t') \tag{11.35}$$

where t and t' denote two points in time. The non-correlation is expressed by the Dirac delta function in the correlation function which features the properties:

$$\delta(\tau) = \begin{cases} 0 & \text{for} \quad \tau \neq 0 \\ \infty & \text{for} \quad \tau = 0 \end{cases} \tag{11.36}$$

and:

$$\int d\tau \, \delta(\tau) = 1 \tag{11.37}$$

and:

$$\int d\tau \, f(\tau)\delta(\tau) = f(0) \tag{11.38}$$

The parameter Γ is a measure for the strength of the fluctuating force and will be determined below. Third, it is often assumed that the fluctuating force features a normal distribution.

We can integrate Eq. (11.33) in terms of the velocity $v = \frac{dx}{dt}$ which then takes the form of a simple first-order differential equation with variable coefficients:

$$\frac{dv(t)}{dt} = k(t)v + l(t) \tag{11.39}$$

which has the general solution:

$$v(t) = v_0 \cdot e^{K(t)} + e^{K(t)} \int_{t_0}^{t} d\tau \, e^{-K(\tau)} l(\tau) \tag{11.40}$$

where we integrate over time expressed by τ, $v(t_0) = v_0$ is an integration constant, and:

$$K(t) = \int_{t_0}^{t} d\tau \, k(\tau) \tag{11.41}$$

Setting $k(t) = \gamma m^{-1}$ and $l(t) = f(t)$ we obtain an expression for the velocity of the molecule as a function of time:

$$v(t) = \frac{dx(t)}{dt} = v_0 e^{-\frac{\gamma}{m}t} + \frac{1}{m} \int_0^t d\tau \, e^{-\frac{\gamma}{m}(t-\tau)} f(\tau) \tag{11.42}$$

where we arbitrarily set $t_0 = 0$. This result enables us to determine $\overline{v^2}(t)$:

$$\overline{v^2}(t) = (v_0)^2 e^{-\frac{2\gamma}{m}t} + \frac{2v_0}{m} \int_0^t d\tau \, e^{-\frac{\gamma}{m}(2t-\tau)} f(\tau)$$
$$+ \frac{1}{m^2} \int_0^t d\tau \int_0^t d\tau' \, e^{-\frac{\gamma}{m}(2t-\tau-\tau')} f(\tau) f(\tau') \tag{11.43}$$

A closer look at this equation reveals that the first term following the equal sign decays quickly and does not contribute to $\overline{v^2}(t)$ on long time scales. Since $\overline{f(t)} = 0$, the second term must vanish, which leaves us only with the third term. Considering the correlation $\overline{f(t)f(t')} = \Gamma\delta(t - t')$ we calculate:

$$\overline{v^2}(t) \approx \frac{1}{m^2} \int_0^t d\tau \int_0^t d\tau' \, e^{-\frac{\gamma}{m}(2t-\tau-\tau')} f(\tau) f(\tau')$$
$$= \frac{\Gamma}{m^2} \int_0^t d\tau \, e^{-\frac{2\gamma}{m}(t-\tau)}$$
$$= \frac{\Gamma}{2m\gamma} (1 - e^{-\frac{2\gamma}{m}t})$$
$$\approx \frac{\Gamma}{2m\gamma} \tag{11.44}$$

where, in the last step, we again acknowledge that the exponential term vanishes quickly with time and can be neglected when $\overline{v^2}(t)$ is calculated for an experimentally relevant time scale. If we now compare this result with the findings (11.32) we made when evaluating the equipartition theorem for the Hamiltonian operator of a free molecule, we can determine the 'strength' of the fluctuating force as defined by the parameter Γ. In the limit of time scales sufficiently long to allow the neglect of the exponentially decaying terms in the above calculations, we find that:

$$\overline{v^2}(t) = \frac{k_B T}{m} = \frac{\Gamma}{2m\gamma} \tag{11.45}$$

and hence:

$$\Gamma = 2\gamma k_B T \tag{11.46}$$

For diffusion in liquids and the time scales relevant to Brownian motion, collisions with solvent molecules are so frequent and the friction force $\gamma\frac{dx}{dt}$ is so large that the acceleration term $\frac{d^2x}{dt^2}$ in the Langevin equation can be neglected. The equation then simplifies to the over-damped Langevin equation:

$$\gamma\frac{dx(t)}{dt} = f(t) \tag{11.47}$$

We can integrate this equation to obtain an expression for the first moment $\overline{x}$ of the distribution of the molecule positions in x. Using the first moment of the fluctuating force, $\overline{f}(t) = 0$, we calculate:

$$\overline{x}(t) = \int_0^{\overline{x}} dx = \frac{1}{\gamma}\int_0^t d\tau\, f(\tau) = 0 \tag{11.48}$$

where, arbitrarily, we set the integration constant in a way that the initial position of the molecule is set to the origin, i.e. $x(t = 0) = 0$. Using our previous result (11.46):

$$\overline{f(t)f(t')} = \Gamma\delta(t - t') = 2\gamma k_B T\delta(t - t') \tag{11.49}$$

for the second moment $\overline{x^2}$ we find:

$$\overline{x^2}(t) = \frac{1}{\gamma^2}\int_0^t d\tau \int_0^t d\tau'\, f(\tau)f(\tau')$$

$$= \frac{1}{\gamma^2}\int_0^t d\tau\, 2\gamma k_B T$$

$$= \frac{2k_B T}{\gamma}t \tag{11.50}$$

where we again set $x(t = 0) = 0$. We note that the second moment of x is proportional to time. Below we return to these important results, $\bar{x}(t) = 0$ and $\overline{x^2}(t) \propto t$.

11.2.1.4. *A General Solution to the Diffusion Equation*

Let us revisit the diffusion equation and find a solution to the hypothetical experiment we investigated in the previous section. In short, we consider a solution of a solvent and a single molecule whose Brownian motion we investigate which, henceforth, is released at the position $x = 0$ at the time $t = 0$. We imagine that we had means to observe its movement in one dimension, that is the projection of the molecule position onto the x-axis, and record a series of the molecules positions $x(t)$ as a function of time. This experiment is repeated several times, aiming to determine the first and second moments of the distribution $p(x, t)$ of particle positions, the expected position $\bar{x}(t)$:

$$\bar{x}(t) = \int dx \, x \, p(x, t) \tag{11.51}$$

and its variance $\overline{x^2}(t)$:

$$\overline{x^2}(t) = \int dx \, x^2 \, p(x, t) \tag{11.52}$$

respectively, as functions of time where $p(x, t)$ follows the definition in Eq. (11.4).

The temporal evolution of the probability density $p(x, t)$ must satisfy the diffusion equation:

$$\frac{\partial p(x, t)}{\partial t} = D \frac{\partial^2 p(x, t)}{\partial x^2} \tag{11.53}$$

subject to the boundary condition:

$$p(x, t = 0) = \delta(x) \tag{11.54}$$

where δ represents the Dirac-delta and is hence a mathematical representation of the fact that the probability of finding the molecule at the time $t = 0$ at the position $x = 0$ equals unity or, in other words, we know that molecule is released at the origin.

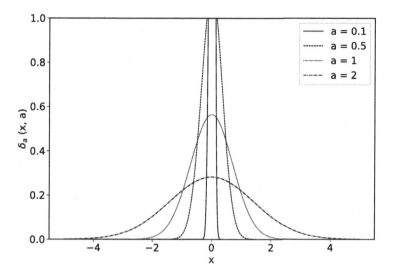

Fig. 11.3. δ_a as a function of x evaluated for different values of a.

The Dirac delta function can be expressed in various ways. In the context of the diffusion equation the following form is often the most practical as it solves diffusion equation (11.53) in the absence of boundary conditions as we will establish below:

$$\delta_a(x) = \frac{1}{a\sqrt{\pi}} e^{-\frac{x^2}{a^2}} \longrightarrow \delta(x) \quad \text{for} \quad a \longrightarrow 0 \tag{11.55}$$

A plot of this function can be seen in Figure 11.3. Assuming a is a function of the time t, we can substitute δ_a in Eq. (11.53) to obtain a differential equation that describes the temporal evolution of a. To this end, we calculate the derivatives:

$$\frac{\partial \delta_a(t)}{\partial x} = -\frac{2x}{(a(t))^3 \sqrt{\pi}} e^{-\frac{x^2}{(a(t))^2}}$$

$$\frac{\partial^2 \delta_a}{\partial x^2} = \left(-\frac{2}{(a(t))^3 \sqrt{\pi}} + \frac{4x^2}{(a(t))^5 \sqrt{\pi}} \right) e^{-\frac{x^2}{(a(t))^2}} \tag{11.56}$$

$$\frac{\partial \delta_a}{\partial t} = \frac{\partial a(t)}{\partial t} \left(-\frac{1}{(a(t))^2 \sqrt{\pi}} + \frac{2x^2}{(a(t))^4 \sqrt{\pi}} \right) e^{-\frac{x^2}{(a(t))^2}}$$

and substitute our results in the diffusion equations (11.53):

$$\frac{\partial a(t)}{\partial t}\left(-\frac{1}{(a(t))^2\sqrt{\pi}}+\frac{2x^2}{(a(t))^4\sqrt{\pi}}\right)e^{-\frac{x^2}{(a(t))^2}}$$

$$= D\left(-\frac{2}{(a(t))^3\sqrt{\pi}}+\frac{4x^2}{(a(t))^5\sqrt{\pi}}\right)e^{-\frac{x^2}{(a(t))^2}}$$

$$\frac{\partial(a(t))}{\partial t}\left(-\frac{1}{(a(t))^2}+\frac{2x^2}{(a(t))^4}\right)$$

$$= D\left(-\frac{2}{(a(t))^3}+\frac{4x^2}{(a(t))^5}\right) \tag{11.57}$$

We now compare the first or the second terms, respectively, on either side of the equation to obtain a differential equation in $a(t)$. In either case we find:

$$\frac{\partial a(t)}{\partial t} = \frac{2D}{a(t)} \tag{11.58}$$

This equation can be solved via separation of variables which yields:

$$\int da(t)\, a(t) = \int dt\, 2D$$

$$\frac{(a(t))^2}{2} = 2Dt + \text{const.} \tag{11.59}$$

$$a(t) = \sqrt{4Dt} + \text{const.}$$

where the integration constant represents a shift in time which, for convenience, we set to zero in all following calculations. $\delta_a(t)$ with $a(t) = \sqrt{4Dt}$ is hence a solution to the diffusion equation and, coming back to our initial hypothetical experiment, we conclude that a probability density $p(x,t)$ of the form $\delta_a(t)$ hence solves the diffusion equation (11.53) subject to the boundary condition (11.54)[4]:

$$p(x,t) = \frac{1}{\sqrt{4\pi Dt}}e^{-\frac{x^2}{4Dt}} \tag{11.60}$$

Since this solution features the structure of a normal distribution, we can straightforwardly determine the first and the second moments, $\overline{x}(t)$ and $\overline{x^2}(t)$, of the probability density which are:

$$\overline{x}(t) = \int dx\, x\, p(x,t) = 0 \tag{11.61}$$

---

[4] For $t \to 0$, $a(t) \to 0$ and $p(x,t \to 0)$ resembles a Dirac delta function as defined in Eq. (11.55), which describes the analyte being released at the position $x = 0$ at the time $t = 0$.

and:

$$\overline{x^2}(t) = \int dx \, x^2 \, p(x,t) = \frac{a^2}{2} = 2Dt \qquad (11.62)$$

In the following subsection, we will come back to these moments $\overline{x}(t)$ and $\overline{x^2}(t)$ for comparison with the results we found via the Langevin equation.

11.2.1.5. *Einstein-Smoluchowski Relation and Stokes-Einstein Equation*

We have established expressions for the first and second moments of the distribution of the positions of a freely diffusing molecule in two different ways, via the Langevin equation of a molecule released at the origin and via the diffusion equation which we have solved for an initial probability density distribution that equals a delta function centred at the molecule's point of release at $x = 0$. In both cases, we have found that the first moments (the expected position of the molecule) is its point of release independently of the time it had to diffuse. Mathematically this was derived via Eq. (11.48) and Eq. (11.61) which state that the first moments are zero. The second moments, however, adopt values different from zero for $t > 0$, which scale linearly with time. In our analysis via the macroscopic diffusion equation we found Eq. (11.50):

$$\overline{x^2}(t) = 2Dt \qquad (11.63)$$

while via the microscopic Langevin equation we calculated:

$$\overline{x^2}(t) = \frac{\Gamma}{\gamma^2}t = \frac{2k_B\mathcal{T}}{\gamma}t \qquad (11.64)$$

We have hence derived two expressions for the second moment, the variance of the displacement that results from the molecule's Brownian motion. These are, on the one hand, functions of the microscopic parameters γ and Γ, the friction coefficient and the 'strength' Γ of the fluctuating force, and on the other, functions of the macroscopic parameters D and $\mathcal{T}$, the diffusion coefficient and the temperature, which can be directly measured from an ensemble of diffusing molecules. The expressions can be combined to yield the Einstein-Smoluchowski relation:

$$\gamma D = k_B\mathcal{T} \qquad (11.65)$$

which is a remarkable achievement as it links friction at the micro scale, temperature, and the macroscopic diffusion coefficient! In the context of this derivation, it provides the crucial link between the diffusion equation and Langevin equation and, equally, between our modelling approach via partial- and via stochastic differential equations.

Substitution of the friction coefficient found in Stokes friction Eq. (11.25), $\gamma = 6\pi r\eta$, gives the commonly-used Stokes-Einstein equation:

$$D = \frac{k_B T}{6\pi r\eta} \tag{11.66}$$

where r is the Stokes radius of the diffusing entity.

11.2.1.6. *Binomial Distributions and the De Moivre-Laplace Theorem*

The previous subsections have shown that the probability density distribution adopts the shape of a normal distribution, which is a continuous function. In a random walk simulation, we however aim to discretise time and space and express the Brownian motion as a sequence of independent stochastic events. The following section will explore links between the normal distribution and related discrete probability distributions.

We consider a Bernoulli process, a sequence of independent stochastic events that can result in either 'success' or 'fail'. This can for instance, represent the repeated flipping of a coin or the path $x(t)$ of a molecule undergoing a Brownian motion in one direction. In the latter case, we can define 'success' as a motion $x \to x + \Delta x$ and 'fail' as a motion $x \to x - \Delta x$. While a Bernoulli processes can describe stochastic processes in which the probabilities of finding 'success' and 'fail' may differ, in the context of Brownian motion where steps are equally likely in both directions, $\bar{x} = 0$, these probabilities are both set to $\frac{1}{2}$, $p('success') = p('fail') = \frac{1}{2}$, which we will use henceforth. The binomial distribution is directly linked to Bernoulli processes as it represents the discrete probability distribution that arises from such repeated 'success'/'fail' experiments. It can be written as:

$$P_n(X = k) = \frac{1}{2^n} \frac{n!}{k!(n-k)!} = \frac{1}{2^n} \binom{n}{k} \tag{11.67}$$

where $n, k \in \mathbb{N}^{\geq 0}$, and n and k represent the number of repeats and the number of times the result 'success' is observed. If, for instance, the experiment is repeated ten times, the probability of obtaining eight times the result 'success' equals $P_{n=10}(X = 8)$. Analogous to the definitions (11.51) and (11.52), we calculate the first moment and the second central moment of $P_n(X = k)$ as:

$$\overline{k} = \sum_{k=0}^{n} k P_n(X = k)$$

$$= \sum_{k=0}^{n} \frac{k}{2^n} \binom{n}{k}$$

$$= \frac{n}{2} \tag{11.68}$$

and:

$$\overline{(k - \overline{k})^2} = \sum_{k=0}^{n} k^2 P_n(X = k)$$

$$= \sum_{k=0}^{n} \frac{k^2}{2^n} \binom{n}{k} - (\overline{k})^2$$

$$= \frac{n}{4} \tag{11.69}$$

The De Moivre-Laplace Theorem establishes the link between the binomial distribution and the normal distribution (11.55):

$$\delta_a(x) = \frac{1}{a\sqrt{\pi}} e^{-\frac{x^2}{a^2}} \tag{11.70}$$

where the parameter a is a finite number. The theorem states that for large n, $n \to \infty$, and a Bernoulli process featuring a probability $0 < p < 1$, the binomial distribution converges to the normal distribution of the form Eq. (11.55). This is illustrated in Figure 11.4 where we plot $\delta_a(x)$ and $P(X = k - \frac{n}{2})$. a is herein chosen in a way that the second central moments, Eq. (11.62) and Eq. (11.69), of both distributions match:

$$\overline{x^2} = \frac{a^2}{2} = \frac{n}{4} = \overline{k^2} \tag{11.71}$$

$$2a^2 = n$$

As can be seen in the Figure, the binomial distribution closely approximates the normal distribution even for small n. The following subsection will exploit this fact and construct a Bernoulli process as a model for Brownian motion.

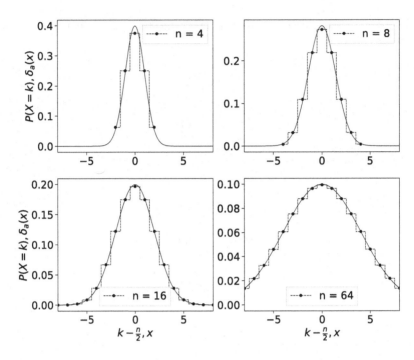

Fig. 11.4. Comparison of the normal distribution Eq. (11.55) (solid line) and the Bernoulli distribution Eq. (11.67) (dashed line) evaluated for different n while the parameter a is set in way that $2a^2 = n$ is satisfied.

11.2.1.7. *Construction of the Random Walk*

Having established that Brownian motion can be represented by a Bernoulli process, that is a sequence of identical stochastic events that each leads to 'success' or 'fail', we can now progress and construct a random walk process on the basis of the equations we have derived throughout this section.

Let us first consider the Bernoulli process depicted on the left of Figure 11.5. On the y axis, we plot the number of identical stochastic events since the beginning of the process which is counted as the integer number n with the difference Δn between two subsequent stochastic events being one. The x axis plots k, the number of times the result 'success' was observed for a given n. Like n, k is an integer number that progresses in steps of Δk equal one. In the plot, we can further see that the expected value $\overline{k}$ increases with n, as we have derived earlier.

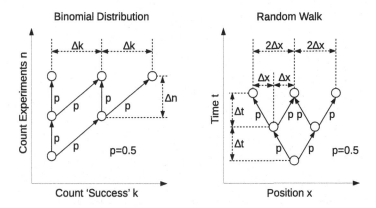

Fig. 11.5. Illustration of a Bernoulli processes that gives rise to the binomial distribution and a random walk.

If we think of this Bernoulli process as a description of the Brownian motion, the discrete n is a measure of time scaled by a factor Δt with:

$$t = n\Delta t \tag{11.72}$$

while the discrete k is a measure of length scaled with a factor which still remains to be determined. Furthermore, the expected position $\bar{x}(t)$, i.e. the first moment of the position of a molecule released at the origin, must vanish. To satisfy this requirement, we define a Bernoulli process as shown on the right of Figure 11.5: Rather than the process on the left in which k either remains constant or increases by one during each step in n, we design the random walk in a way that, at each time step, the molecule can either perform a 'forward' step of Δx or a 'backward' step of $-\Delta x$. This necessarily satisfies the requirement of a vanishing expected value $\bar{x}(t)$ as we have set the probabilities for steps in either direction to $\frac{1}{2}$. The distance between the two positions the random walker may take after each step is then $2\Delta x$ on the right instead of Δk on the left.

To determine the value of Δx we compare the second moments of the binomial distribution and our corresponding solution (11.60) to the diffusion equation:

$$\overline{(k - \bar{k})^2} = \frac{n}{4} \quad \text{and} \quad \overline{x^2}(t) = 2Dt \tag{11.73}$$

To this end, we multiply both sides of the definition of $\overline{(k - \bar{k})^2}$ by $8D\Delta t$ to reproduce the right side of the definition of $\overline{x^2}(t)$ before we combine the remaining factors to $2\Delta t$:

$$\overline{(k - \overline{k})^2} = \frac{n}{4} \qquad \Big| \cdot 8D\Delta t$$

$$8D\Delta t \cdot \overline{(k - \overline{k})^2} = 2Dn\Delta t$$

$$\underbrace{2^2(\sqrt{2D\Delta t})^2}_{=(2\Delta x)^2} \cdot \overline{(k - \overline{k})^2} = 2Dn\Delta t \qquad (11.74)$$

$$\overline{(2\Delta x)^2 (k - \overline{k})^2} = 2Dn\Delta t$$

$$\overline{x^2}(t) = 2Dt$$

This comparison yields the scaling factor in space which is:

$$\Delta x = \sqrt{2D\Delta t} \qquad (11.75)$$

In summary, we conclude that the Brownian motion of a molecule can be modelled via a simple Bernoulli process: During each time step Δt the molecule is displaced by $\Delta x = \sqrt{2D\Delta t}$. Figure 11.6 compares the results of 1000 of such random walks with the corresponding analytical solution (11.60). As can be seen in the figure, on average, the random walk results resemble the analytical distribution while the stochasticity of the process can be seen in the fluctuations of the probability density functions.

11.2.1.8. *Random Walks in Two and Three Dimensions*

Our derivation of the random walk can be extended to describe Brownian motion in two and three dimensions. To this end, we write the Langevin equation in absence of external forces and in three dimensions:

$$m\frac{d^2\vec{r}(t)}{dt^2} = -\gamma\frac{d\vec{r}(t)}{dt} + \vec{f}(t) \qquad (11.76)$$

This equation can equally be expressed as three independent stochastic differential equations in x, y, and z:

$$m\frac{d^2x(t)}{dt^2}\vec{e_x} = -\gamma\frac{dx(t)}{dt}\vec{e_x} + f_x(t)\vec{e_x}$$

$$m\frac{d^2y(t)}{dt^2}\vec{e_y} = -\gamma\frac{dy(t)}{dt}\vec{e_y} + f_y(t)\vec{e_y} \qquad (11.77)$$

$$m\frac{d^2z(t)}{dt^2}\vec{e_z} = -\gamma\frac{dz(t)}{dt}\vec{e_z} + f_z(t)\vec{e_z}$$

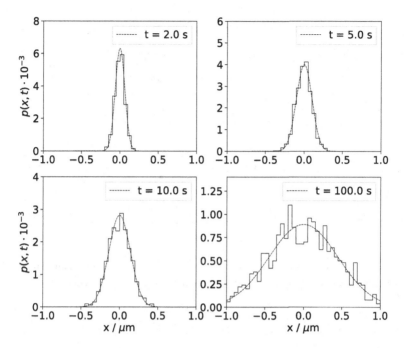

Fig. 11.6. Comparison between the analytical result (11.60) and data obtained from 1000 random walks modelled on the basis of the equations (11.72) and (11.75). The diffusion coefficient is set to 10^{-9} m^2 s^{-1} and Δt is set to 1 ms.

all of which feature the same structure as our solution in one dimension. The Brownian motion in two or three dimensions can hence be modelled as two or three one-dimensional random walks, respectively, with:

$$\Delta x = \Delta y = \Delta z = \sqrt{2D\Delta t} \tag{11.78}$$

where the random walks for each direction are independent of each other and progress simultaneously with the time steps Δt.

Figure 11.7 illustrates two- and three-dimensional composed of independent one-dimensional random walks. In the figure, we can see that, in each time step of a two-dimensional random walk, there are four different positions $((x + \Delta x, y + \Delta x)$, $(x - \Delta x, y + \Delta x)$, $(x + \Delta x, y - \Delta x)$, and $(x - \Delta x, y - \Delta x))$ to which the analyte can move while the overall step width dx' is given by:

$$dx' = \sqrt{dx^2 + dx^2} = \sqrt{4D\Delta t} \tag{11.79}$$

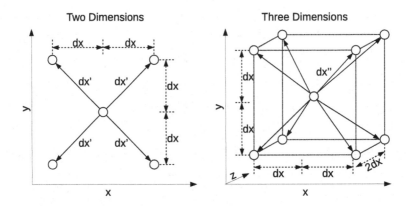

Fig. 11.7.   Illustration of random walk simulations in two and three dimensions.

as $dx = \sqrt{2D\Delta t}$. In three dimensions, we analogously find that there are eight directions in which the analyte can move and the overall step width dx'' is:

$$dx'' = \sqrt{dx^2 + dx^2 + dx^2} = \sqrt{6D\Delta t} \tag{11.80}$$

11.2.2. *Example: Reversible Voltammetry at a Macroelectrode*

In the following example, we revisit the electrochemical system we discussed in Chapter 3: A Nernstian reaction at a macro-electrode is investigated via cyclic voltammetry. While previously we have established a computational model by means of finite differences, we now turn to random walks and make use of the knowledge we have gained in this chapter.

We consider the reduction of a species A to B via a one-electron transfer reaction of the form:

$$A + e \rightleftharpoons B \tag{11.81}$$

The experiment is carried out in a solution featuring excess amounts of electrolyte to ensure that all electric field components that originate from the electrode are screened throughout the solution. The mass transport of both species hence follows the linear one-dimensional diffusion equation:

$$\frac{\partial c_A}{\partial t} = D\frac{\partial^2 c_A}{\partial x^2} \quad \text{and} \quad \frac{\partial c_B}{\partial t} = D\frac{\partial^2 c_B}{\partial x^2} \tag{11.82}$$

where we note that both species feature equal diffusion coefficients $D_A = D_B = D$. The reaction at the interface is Nernstian and a thermodynamic

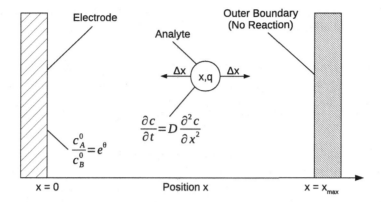

Fig. 11.8. Schematic illustration of the theoretical model used in the random walk example. The state of each molecule is fully characterised by its position x and the charge q it carries.

equilibrium is hence established at all time. The ratio of the surface concentrations is then a function of the electrode potential and given by the Nernst equation:

$$\frac{c_A^0}{c_B^0} = exp\left(\frac{F}{\mathcal{R}\mathcal{T}}(E - E_f^0)\right) = e^\theta \tag{11.83}$$

where c_A^0 and c_B^0 denote the concentrations of A and B, respectively, at the electrode surface at $x = 0$. A schematic illustration of the investigated system can be seen in Figure 11.8.

At the beginning of the experiment and before the potential sweep begins, the electrode is held at a sufficiently high oxidising potential to ensure that only species A is present in the solution. The concentration of A is further assumed homogeneous at c^*, which yields the boundary conditions:

$$c_A = c^* \qquad \text{for } t = 0 \text{ and all } x \tag{11.84}$$

$$c_B = 0 \qquad \text{for } t = 0 \text{ and all } x \tag{11.85}$$

in all space and at the time $t = 0$, which is the point in time when the potential sweep is initiated.

The diffusion of A and B is modelled via random walks which feature a temporal step width of Δt and a spatial step width of Δx. Following our above derivation, the step widths are linked via Eq. (11.75):

$$\Delta x = \sqrt{2D\Delta t} \tag{11.86}$$

At each time step, each random walker is displaced by either plus or minus Δx at a probability of 50%:

$$x \xrightarrow{p=\frac{1}{2}} x + \Delta x \quad \text{or} \quad x \xrightarrow{p=\frac{1}{2}} x - \Delta x \tag{11.87}$$

If an analyte features a position $x < 0$, it has reached the electrode and its path is reflected at the electrode surface:

$$\text{if } x < 0 \text{ then } x \longrightarrow -x \tag{11.88}$$

In addition and upon each contact with the electrode, we further re-evaluate the oxidation state of the respective analyte. If we consider a large ensemble of analytes at the electrode surface and measure the oxidation state of a random analyte, the probability p_A of finding it in the oxidation state of species A is determined by the surface concentrations of A and B:

$$p_A = \frac{c_A^0}{c_A^0 + c_B^0} \tag{11.89}$$

Substitution of the Nernst equation (11.83), $c_A^0 = c_B^0 e^\theta$, then yields:

$$p_A = \frac{e^\theta}{1 + e^\theta} = \frac{1}{1 + e^{-\theta}} \tag{11.90}$$

and analogously for the probability of finding the analyte in the oxidation state of B:

$$p_B = \frac{1}{1 + e^\theta} = 1 - p_A \tag{11.91}$$

Upon each contact of an analyte with the electrode, we can hence randomise its oxidation state using p_A and p_B to model a Nernstian equilibrium. If, in such a reaction, the oxidation state is changed and A turns into B or *vice versa*, the reaction contributes $\pm \Delta I$ to the electrode current:

$$\Delta I = \pm \frac{e_0}{\Delta t} \tag{11.92}$$

where the plus and the minus signs represent oxidations of B to A and reductions of A to B, respectively.

The outer boundary is set according to the equations (2.20) and (11.11):

$$x_{max} = 6\sqrt{Dt_{max}} \tag{11.93}$$

and, akin to the electrode, the analyte path is reflected if $x > x_{max}$:

$$\text{if } x > x_{max} \text{ then } x \longrightarrow 2x_{max} - x \tag{11.94}$$

while there is no reaction and its oxidation state remains unchanged.

11.2.2.1. *Implementation in C++*

In the following section, we discuss the implementation of our random walk example of a reversible reaction at a macro-electrode. To this end, we begin with some practical considerations that are applicable to random walk simulations in general before we explore the details of an implementation in C++.

At the beginning of any new simulation project, detailed planning is crucial. This especially holds true for the development of stochastic simulations, where, depending on the problem attacked, a very significant computational workload may arise. In the design of a random walk simulation, there are five key decisions that have to be made ahead of the actual code development:

(1) **Should I use a random walk approach?**

While at first glance and often falsely, the implementation of a random walk may appear easier than a finite differences model, it is paramount to be aware of the limitations of random walks. Random walk simulations allow insight into fluctuations in the electrode current and the microscopic state of an electrochemical system, but fail in the modelling of systems in which the electrode current reflects a large number of individual analyte reactions within each sampling interval and when the computational workload is too large to be computed within a helpful amount of time. If there is no interest in the stochastics of a system or if fluctuations average out, random walk simulations should be avoided and finite differences should be used.

(2) **Is a random walk approach feasible?**

In contrast to the finite-difference simulations we have discussed throughout most of this book, random walk simulations usually require the computation of a very significant workload, which may be prohibitively large. It is therefore paramount to estimate the time of computation that is required for the planned simulation on the available computer hardware.

Since the number of individual random walk steps that must be calculated scales linearly with the duration of the simulated experiment $N_{steps} = \frac{t_{max}}{\Delta t}$, the number of modelled analytes $N_{analyte}$, and the degrees of freedom N_{dim}, the required computation time t_{comp} can be estimated on the basis of a simple test simulation in which the computation time t_{step} of an individual random walk step is measured:

$$t_{comp} \approx N_{steps} N_{analyte} N_{dim} t_{step} \tag{11.95}$$

which is the estimated computation time if no random walks are calculated in parallel.

(3) **Which coding language should I choose?**
In random walk simulations, performance is always the one argument that should determine the choice of the coding language and the compiler. While opportunities in various languages evolve at a rapid pace, C++ is currently an excellent starting point for the development of most random walk simulations.

(4) **Which pseudo random number generator should I choose?**
Random walk simulations crucially rely on the pseudorandom number generator (PRNG) that is used to model the stochastics of the Brownian motion. These number generators mimic a particular stochastic process on the basis of a seed number while identical seed numbers lead to identical sequences of pseudorandom numbers. Stochastic processes are always modelled with limited accuracy while there is a trade off between accuracy and performance which must be considered in the choice of the PRNG. Another important feature to consider is the length of the sequence of random numbers that can be generated: If more numbers are calculated than the length of the sequence of the PRNG, the sequence may either repeat or otherwise fail to resemble the desired stochastic process. Older implementations of the commonly used *std::rand()* in C++ may for instance only offer relatively short periods while PRNGs based on the Mersenne Twister [7] are often suitable for large random walk simulations.

(5) **Can I use parallel computing techniques?**
The use of parallel computing techniques in random walk simulations is hugely advantageous. In all random walk simulations discussed in this book, the individual random walk steps are entirely independent of each other in time and space, and hence bear an enormous potential for parallel computing: The computational workload of the random walk simulation can be easily divided between different computing units, which leads to a significant increase in the computational performance.

The opportunities for a parallel implementation in terms of coding tools are widespread and rapidly evolving. In C++, OpenMP is an application programming interface that enables multi-processing on shared-memory systems. This can, for instance, be a workstation or a laptop featuring a multi-core CPU with the individual CPU cores where the cores share the same random access memory (RAM). Given that many current CPUs in standard computers are such multi-cores CPUs, the use of OpenMP can result in a significant increase in the performance of random walk simulations without the need to acquire new hardware. Even greater performance gains can currently be achieved via GPUs that often feature a significantly greater number of cores for parallel computation.[5] GPUs can, for instance, be used via CUDA or OpenCL which may however be less accessible to beginners in software development.

We now return to the implementation of our example of reversible voltammetry at a macro-electrode and establish an implementation in C++. As in all other code examples in this book, our discussion aims to provide a starting point for the development of more sophisticated simulations rather than a perfect implementation. The code is deliberately kept simple to illustrate an almost minimal working example and to avoid complexity, which arises from more advanced coding techniques like exception handling, parallel computing, or similar ones. In contrast to previous code examples, the following one makes greater use of object oriented programming. Please note that the code uses features defined in the C++11 ISO. The inclusion of respective compiler flags, such as '-std=c++11' in g++, may hence be required for the compilation.

Figure 11.9 provides a schematic illustration of the simulation procedure. First, parameters are set and dependent parameters, such as the number of simulated molecules being a function of the analyte concentration and the simulated volume, are determined. Second, time (and hence space) are discretised via the definition of discrete time steps and memory is allocated to collect data. The third step is repeated for each modelled molecule: We generate an A molecule (see Eq. (11.81)) at a random position in x to match the boundary conditions (11.84) and (11.85). The random walk is then executed while the current contributions of the molecule to the overall current are collected before the molecule is deleted and we progress

---

[5] At the time of writing, recent multi-core CPUs offer 12 cores while recent GPUs feature 5000 cores.

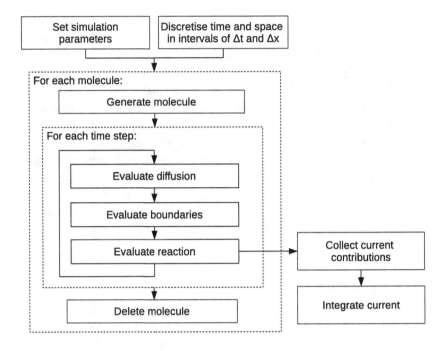

Fig. 11.9.   Flow diagram of the simulation procedure followed in the example code.

to the next random walk. Fourth and finally, the current contributions are combined and integrated to a sampling rate that matches the sampling rate of the modelled experiment.

The simulation code commences with the inclusion of various libraries. We link the `fstream` library which will later allow us to use the `std::ofstream` class object, a simple tool to save data in text files. In addition, we include `math.h` and the general utilities of the C Standard Library `stdlib.h` to enable simple mathematical operations and access to a pseudorandom number generator, and set the namespace to `std` for convenience and readability of the code.

```
#include <fstream>
#include <math.h>
#include <stdlib.h>
using namespace std;
```

In the next step, various constants are defined which will be used throughout the entire code. The values are stored in a **struct** which enables convenient handling via a single name and ensures that the data is only available where it is actually needed. We will later generate a `const` instance of this data structure to ensure that it cannot be amended inadvertently. The definition of global constants would, of course, be an equally adequate approach.

```
struct Constants
{
    double NA = 6.022e23;    // Avogadro's number [mol^-1]
    double e0 = 1.602e-19;   // elementary charge [C]
    double F = 96485.33;     // Faraday constant [s A mol^-1]
    double R = 8.314;        // universal gas constant
                             // [J K^-1 mol^-1]
    double T = 298.15;       // temperature [K]
                             // approx 25 C
};
```

In the next step, we define a `class` that contains all simulation parameters as well as a function `void evalImplicitParameters(Constants c)` to calculate the values parameters that are derived from the simulation parameters and the above constants. Since `Parameters` is a `class` object, we define the constructor `Parameters()` and the destructor `~Parameters()` though they remain unused in this example.

It is important to note that, for simplicity, our simulation describes only linear- rather than radial diffusion. Since we solve the one-dimensional linear diffusion equation for a nano-scale electrode (see the definition of the electrode area in the code below), we model a geometry in which the analyte is confined in a way that it can only reach the electrode from the direction perpendicular to the electrode surface. This is for instance the case for a nano-electrode at the bottom of a long cylindrical pore. To model diffusion towards a nano-band or nano-disk electrode, the code must be extended to two- and three-dimensional orthogonal random walks, respectively.

```
class Parameters
{
    public:
    // scan rate [V s^-1]
    double v = 1.0;

    // target concentration before discretisation [mM]
    double c0 = 1.0e-1;
```

```
// electrode surface area [m^2]
double A = 3.142 * 100.0e-9 * 100.0e-9;

// formal potential [V]
double EF = 0.0;

// min electrode potential [V]
double EMin = -.200;

// max electrode potential [V]
double EMax = .200;

// diffusion coefficient [m^2 s^-1]
double D = 1e-9;

// sampling rate potentiostat [Hz]
double fs = 500.0;

// oversampling in the random walk
// (f_rw = nOversampling * f_s)
int nOversampling = 100;

// length of the simulated time interval [s]
double tMax;

// width of the simulated space [m]
double xMax;

// number of molecules
int nMolecules;

// number of time steps
int nTimeSteps;

// random walk temporal step length [s]
double dt;

// constructor
Parameters()
```

```
{
    // nothing to do here
}

// destructor
~Parameters()
{
    // nothing to do here
}

void evalImplicitParameters(Constants c)
{
    tMax = 2.0 * (EMax - EMin) / v;
    xMax = 6.0 * sqrt(D * tMax);
    nMolecules = int(c.NA * c0 * xMax * A);
    nTimeSteps = int(tMax * fs * nOversampling);
    dt = 1.0 / fs / nOversampling;
}
};
```

The following defines the pseudorandom number generator that is used throughout the simulation. In the example, we base the pseudorandom number generation on `std::rand()`, which, as discussed above, may feature only a relatively short period. While this is sufficient for the purpose of this illustration, more advanced generators[6] should be used for larger projects. `std::rand()` generates an integral random number between 0 and RAND_MAX. Casting both into double precision numbers and dividing `std::rand()` by RAND_MAX hence calculates a pseudorandom double precision number between 0 and 1.

```
double rnd()
{
    // return random number between 0 and 1
    return (double(rand()) / (double(RAND_MAX)));
};
```

---

[6] The `<random>` and the `gsl/gsl_rng.h` header in the C++ Standard Library and the GNU Scientific Library, respectively, provide powerful pseudorandom number generators for a wide range of purposes.

The `class Voltammogram` comprises a number of features that are required to record a cyclic voltammogram. In the class constructor, memory is allocated to store the points in time `t[i]` at which the discretised time steps take place, the corresponding electrode potentials `E[i]`, and the corresponding electrode currents `I[i]`. The arrays `t[i]` and `E[i]` are further evaluated while `I[i]` is initialised with zeros. The class further includes a function `void averageAndSave()`, which is of great importance: Random walks are usually evaluated for very small temporal step widths Δt that are much smaller than the duration of a sampling interval of the potentiostat. If, for instance, a potentiostat samples a current at 100 Hz, a data point is measured every $\Delta t_s = \frac{1}{100\ Hz} = 10\ ms$ while a typical random walk may feature a Δt in the low μs range. During such short time steps Δt in the random walk, no or only few molecules may react and simulated data may rather be a series of spikes than the voltammogram seen experimentally. It is therefore essential to reduce the sampling rate of the modelled current to enable a direct comparison with the experimental data. Assuming that there are no bandwidth limitations in the experimentally found current and that the potentiostat records the current throughout the entire sampling interval, this is achieved by calculating the average current over multiple time steps whose durations Δt sum up to the length of the sampling interval Δt_s of the potentiostat.

```
class Voltammogram
{
public:
    double* t;      // time series
    double* E;      // electrode potential series
    double* I;      // electrode current series
    Parameters p;   // class-internal copy of parameters

    // constructor
    Voltammogram(Parameters p)
    {
        // save parameters
        this->p = p;

        // allocate memory
        t = new double[p.nTimeSteps];
        E = new double[p.nTimeSteps];
        I = new double[p.nTimeSteps];
```

```
    // initialise lists
    for (int i=0; i<p.nTimeSteps; i++)
    {
        // set time steps
        t[i] = p.dt*i;

        // set potential sweep
        if (i < p.nTimeSteps/2.0)
        {
            E[i] = p.EMax - p.v * p.dt * i;
        }
        else
        {
            E[i] = p.EMin + p.v * p.dt
                * (i - p.nTimeSteps/2.0);
        }

        // initialise current list
        I[i] = 0.0;
    }
}


// destructor
~Voltammogram()
{
    // free memory
    delete[] t;
    delete[] E;
    delete[] I;
}


void averageAndSave()
{
    // create file
    ofstream file;
    file.open("data.txt");

    // integrate and save
    double aux = 0.0;
```

```
for (int i=1; i<p.nTimeSteps; i++)
{
    aux += I[i];
    if ((i)%p.nOversampling == 0.0)
    {
        file << t[i] << "    "
             << E[i] << "    "
             << aux / p.nOversampling
             << endl;
        aux = 0.0;
    }
}

// close file
file.close();
}
};
```

The class `Molecule` gathers the data required to describe the state of each molecule (position and charge), a function to execute individual random walk steps, and a function to evaluate the molecule's interaction with the electrode. It is herein important to note that the function `bool move(bool up)`, which executes a random walk step, returns a bool `true` if the molecule reaches the electrode surface as a result of the random walk step. This bool value is later used to trigger the evaluation of the electrode interaction via the function `double evalChargeTransfer(double E, double r)`, which then returns the molecule's contribution to the overall current $\pm \frac{e_0}{\Delta t}$ during the respective step if a reaction takes place.

```
class Molecule
{
private:
    double position;    // position [m]
    bool oxidised;      // oxidation state,
                        // true=oxidised, false=reduced
    double dx;          // random walk step width
    Parameters p;       // class-internal copy of parameters
    Constants c;        // class-internal copy of constants

public:
```

```
// constructor
Molecule(double xInit, Parameters p, Constants c)
{
    // set member variables
    oxidised = true;
    position = xInit * p.xMax;
    this->p = p;
    this->c = c;

    // evaluate random walk displacement
    dx = sqrt(2.0 * p.D * p.dt);
}

// destructor
~Molecule()
{
    // nothing to do
}

bool move(bool up)
{
    // evaluate random walk.
    // return true if contact with electrode

    // random walk
    if (up)
    {
        position += dx;
    }
    else
    {
        position -= dx;
    }

    // check electrode boundary
    if (position <= 0.0)
    {
        position = - position;
        return true;
```

```
    }

    // check upper boundary
    if (position >= p.xMax)
    {
        position = 2.0 * p.xMax - position;
    }

    return false;
}

double evalChargeTransfer(double E, double r)
{
    // calculate probability of
    // the molecule being oxidised
    double p_ox = exp(c.F / c.R / c.T * E)
        / (1.0 + exp(c.F / c.R / c.T * E));

    // initialise aux variable
    double currentContrib = 0.0;

    // evaluate current
    if (r < p_ox)
    {
        if (!oxidised)
            currentContrib = c.e0 / p.dt;
        oxidised = true;
    }
    else
    {
        if (oxidised)
            currentContrib = - c.e0 / p.dt;
        oxidised = false;
    }

    // return result
    return currentContrib;
    }
};
```

We finally turn towards the `int main()` function of the programme. Here, we initially create instances of various structures and classes: The `struct Constants`, which is set `const` to ensure values are not unintentionally altered during the runtime of the programme, the `class Parameters` alongside a call of the function `void Parameters::evalImplicitParameters(Constants c)` to evaluate dependent parameter values, and the `class Voltammogram`. We then evaluate the individual random walks molecule by molecule and time step by time step as implemented in the nested `for` loops in `n` and `i`. After the final random walk, the simulated current is evaluated and saved via the function `void Voltammogram::averageAndSave()`.

```
int main()
{
    // generate instance of Constants
    const Constants c;

    // generate instance of Parameters
    Parameters p;
    p.evalImplicitParameters(c);

    // generate instance of Voltammogram
    Voltammogram v(p);

    // calculate random walks
    for(int n=0; n<p.nMolecules; n++)
    {
        // initialise molecule
        Molecule m(rnd(), p, c);

        // calculate path
        for (int i=0; i<p.nTimeSteps; i++)
        {
            bool evalChargeTransfer
                = m.move(bool (round(rnd())));
            if (evalChargeTransfer)
                v.I[i] += m.evalChargeTransfer(v.E[i], rnd());
        }
    }
```

Fig. 11.10. Example voltammogram calculated with the code introduced in this section. Simulated data is further compared with the Randles-Ševčík equation and the theoretical peak positions for the oxidative and the reductive peak. The simulation models an analyte concentration of 100μM, a diffusion coefficient of 10^{-9} m^2 s^{-1}, an electrode radius of 100 nm, a scan rate of 1 V s^{-1}, a potentiostat sampling rate of 500 Hz, and a random walk sampling rate of 50 kHz.

```
// save voltammogram in file
v.averageAndSave();

// end
return 0;
}
```

Figure 11.10 depicts an example voltammogram calculated with the above code. As can be seen in the figure, the intrinsic fluctuations of the electrode that arise from the stochastic mass transport and reaction of the analyte are reflected in the voltammogram. In addition, the figure compares simulated data with analytical theory for the height of the reductive peak and both peak positions, which illustrates one of the complexities of stochastic simulations in terms of their validation: In contrast to finite differences simulations, which in limiting cases can directly be compared with analytical theory, results of stochastic simulations are generally affected by fluctuations which may mask potential errors in the programme or a potential lack of convergence in Δt. Greater confidence can only be reached via the repetition and subsequent averaging of multiple simulation results for the comparison with analytical theory while different seeds are used for the pseudorandom number generators.

References

[1] C. Batchelor-McAuley, J. Ellison, K. Tschulik, P. L. Hurst, R. Boldt, and R. G. Compton, In situ nanoparticle sizing with zeptomole sensitivity, *Analyst.* **140**, 5048–5054, (2015).

[2] P. S. Singh and S. G. Lemay, Stochastic Processes in Electrochemistry, *Anal. Chem.* **88**, 5017–5027, (2016).

[3] S. V. Sokolov, S. Eloul, E. Kätelhön, C. Batchelor-McAuley, and R. G. Compton, Electrode–particle impacts: A users guide, *Phys. Chem. Chem. Phys.* **19**, 28–43, (2017).

[4] M. A. G. Zevenbergen, P. S. Singh, E. D. Goluch, B. L. Wolfrum, and S. G. Lemay, Electrochemical Correlation Spectroscopy in Nanofluidic Cavities, *Anal. Chem.* **81**, 8203–8212, (2009).

[5] W. Cheng, X.-F. Zhou, and R. G. Compton, Electrochemical Sizing of Organic Nanoparticles, *Angew. Chem. Int. Edit.* **52**, 12980–12982, (2013).

[6] L. Sepunaru, B. J. Plowman, S. V. Sokolov, N. P. Young, and R. G. Compton, Rapid electrochemical detection of single influenza viruses tagged with silver nanoparticles, *Chem. Sci.* **7**, 3892–3899, (2016).

[7] M. Matsumoto and T. Nishimura, Mersenne twister: A 623-dimensionally equidistributed uniform pseudo-random number generator, *ACM T. Model. Comput. S.* **8**, 3–30, (1998).

Appendix A: Review of C++

Here we will briefly introduce some of the essential C++ concepts that are used in programs throughout this book. This is by no means intended to provide comprehensive coverage of the language, but should allow those familiar with general programming concepts but with no C++ experience to write simple C++ programs, or at the very least translate the programs and program fragments in this book in their language of choice. The language constructs utilised in this work are found, with little variation, in most other modern programming languages so such translation should be straightforward in most cases.

We will cover only the basic features of C++ that are required to understand the code contained in this book. Most of the more advanced features of the language will not be covered such as pointers, exceptions and templates, and the concepts related to object-oriented programming such as classes, inheritance and polymorphism. Such features are invaluable when creating larger, general-purpose simulation packages but were not felt to be necessary for understanding and developing the (structurally) simple programs covered in this text. A very large number of textbooks are available for learning C++ and on more advanced topics [1–7].

Since the aim here is to develop simple electrochemical simulations, we will forgo any discussion of graphical user interfaces as this would introduce unnecessary complexity, and concentrate solely on developing command-line-interface-based programs. There are a wide variety of C++ compilers available for all operating systems and you should refer to the instructions provided with your compiler for the specific compilation procedure.

A.1. Hello World

In C++ a typical "Hello world" program is written as

```
#include <iostream>
```

```
using namespace std;

// A C++ program always begins at the main() function
int main()
{
    cout << "Hello World!";

    return 0;
}
```

We will now examine this code line by line. The first line in the program is

```
#include <iostream>
```

This is an example of an `#include` directive, which is used to import C++ code that is found in other files into the current file. In this case, we are including the *header* file `iostream` which is part of the C++ standard library and contains functionality for input and output. It is necessary to include this header if you want to use this functionality. The next line is

```
using namespace std;
```

The `std` namespace is the namespace in which the entire C++ standard library is declared. Including the above line in your program introduces it to all of the names in the standard library (which includes `cout`) and is therefore necessary for successful compilation. Further discussion of namespaces is beyond the scope of this introduction.

The next line is an example of a single-line comment. In C++, a comment begins with `//` and extends to the end of that line; everything after the `//` symbol is completely ignored by the compiler. This allows for executable code and a comment in the same line, as long as the executable code comes first. C++ also allows for multi-line comments which begin with `/*`, and end with `*/` and can be spread over any number of lines; everything between these two symbols is ignored by the compiler.

The remainder of the program defines the actual executable code. The line

```
int main()
```

defines a function (subroutine), `main()`, that takes no parameters and returns a value of integer type. All C++ programs consist of one or more functions (discussed in more detail shortly), but all programs must include a function called **main()** as this is where program execution begins by default. The definition of **main()**, a series of executable *statements* that describes what it does, is enclosed in a matched pair of curly braces, { and }. The statement

```
cout << "Hello World!";
```

causes the message "**Hello World!**" to appear on the console. The operator **<<** causes whatever appears to the right-hand side of it to be sent to the object on the left-hand side, in this case, **cout**, which is an identifier that stands for console output. Note that multiple messages can be sent to the console by a single line of code, by chaining multiple **<<** together, i.e.,

```
cout << "Message 1 " << "Message 2";
```

The final statement, "**return 0**" terminates the main function and therefore the program, and returns a value of 0 to the operating system, indicating that the program is terminating normally. Notice that all statements must end with a semi-colon.

A.2. Variables, Types and Operators

Generally speaking, a variable is a named location in memory that can be assigned a value. In C++, a variable name must be declared before a value can be assigned to it:

```
int size;
size = 4;
```

Here the first statement declares a variable of type **int** (an integer) called **size**; the second statement assigns the value 4 to this variable. Note that the declaration may also appear on the same line as the initial assignment:

```
int size = 4;
```

In addition to the integer, the fundamental types in C++ are single- and double-precision floating point (**float** and **double** respectively), boolean (**bool**) and character (**char**). Variables of the floating point types can hold numerical values with fractional components as well as very large numbers that are too big to be stored in a variable of type **int**:

```
float frac, big;
frac = 7.89;
big = 7.24E12;
```

A `double` can hold numbers to a higher precision (more significant figures) than can a `float`. A variable of type `bool` can hold the value `true` or the value `false`, both of which are keywords. Additionally, in C++, numerical values automatically evaluate to true or false, with any non-zero value evaluating to true and zero evaluating to false. A variable of type `char` can hold a single 8-bit ASCII character:

```
char letter = 'z';
```

The ASCII character set includes the letters a–z, A–Z, numbers 0–9, and most common punctuation marks. ASCII also includes a number of control characters, the most important of which is written as \n, which is the code for a new line. Unlike many other languages, there is no fundamental string type; however, one is included as part of the standard library.

In C++, mathematical operations are performed through use of the following binary operators: * (multiplication), / (division), + (addition), and – (subtraction). Multiplication and division have a higher precedence than addition and subtraction, though this can be overridden through the use of parentheses, (); operations of equal precedence are carried out left to right. The following are examples of integer operations:

```
int a, b;
a = 4 + 3 * 5;      // a has value '19'
a = (4 + 3) * 5;    // a has value '35'
b = a - 21;         // b has value '14'
a = b / 4           // a has value '3'
```

Note that the result of integer division (as in the last line above) is rounded down. To find the remainder from such a division, the modulus operator, %, is used; for example, the result of the operation 14 % 4, is 2. The operators ++, and -- can be used as a shorthand for increment and decrement respectively. For example, the statement "a++;", is equivalent to the statement "a = a + 1;". Further, there exists a set of compound assignment operators which take the form *op*=. As a demonstration of their usage, the following two statements are equivalent:

```
a = a + 10;
a += 10;
```

A.2.1. *Arrays, vectors, and strings*

An array is a collection of variables of the same type that are referred to by a common name and may be declared and accessed as in the following example:

```
int heights[5];
heights[0] = 18;
heights[4] = 27;
heights[5] = 19;    // error - index too large
```

In C++, arrays are zero-indexed; they have 0 as the index of the first element. Consequently, as in the above example, the index of the last element of an array of size n is $n - 1$; attempting to access the elements at index n is an error, but C++ does not automatically detect when this occurs, which can lead to significant problems as other sections of memory may be overwritten. For this reason it is common to use another structure called a `vector` which is a type of array included as part of the C++ standard library which can be dynamically resized and can also check whether the element that you are attempting to access exists. To use vectors in your program you must include the appropriate header using the include directive:

```
#include <vector>
```

A vector is declared and used in the following way:

```
vector<int> lengths;
lengths.push_back(27);
lengths.push_back(12);
lengths.push_back(15);

cout << "Vector has" << lengths.size() << "elements\n";
cout << "The first element is" << lengths[0];
cout << "The last element is" << lengths.back();
```

The declaration of a vector must include the type of the elements that the vector will store in angled brackets. It is not necessary to specify the size of vector when it is declared; rather, the vector starts out empty and new elements are appended to the back of the vector with the `push_back()` member function. The current number of elements in the vector is accessed using the `size()` member function; the value of the last element (most recently added) may be accessed with the `back()` member function. Vectors also support random read/write access by index using the `[]` notation.

A string is an object that can hold a sequence of characters. One way of implementing such an object is to use an array of type `char`; however, the C++ standard library includes a `string` type which is more convenient. To use strings in your program, you must include the header with the include directive: `#include <string>`. Strings are used in the same manner as the other types (`int`, `char`, etc.). The following outputs the message "`This is a message!`" to the console.

```
string msg1 = "This is a";
string msg2 = " message!";
cout << msg1 + msg2;
```

Note that strings must be enclosed in double quotation marks whereas characters were enclosed in single quotation marks. Two strings may be concatenated (joined together) using the + operator.

A.3. Logic and Program Control

A.3.1. *The if statement*

The `if` statement allows for conditional execution of code by executing a code block only if a specified condition is determined to be `true`:

```
if(x > 10)
{
    cout << "x is greater than 10";
}
else
{
    cout << "x is less than 10";
}
```

The condition in parentheses can be any expression that evaluates to `true` or `false` (remembering that numerical values are automatically converted to true/false). For this purpose we can make use of the *relational operators*: `>` (greater than), `<` (less than), `>=` (greater than or equal to), `<=` (less than or equal to), `==` (equal to), and `!=` (not equal to). For example, the expressions $(5.4 > 1.8)$, $(-4 <= 12)$, and $(8.2 != 0.46)$ all evaluate to `true`, whereas $(7 > 9)$, $(12 == 3)$ and $(3 != 3)$ all evaluate to `false`.

We may also make use of the standard logical operators: `&&` (and), `||` (or) and `!` (not). The expression (*operand* `&&` *operand*) will evaluate to `true` if both operands are themselves equal to `true` but will evaluate to `false` otherwise. The expression (*operand* `||` *operand*) will evaluate to `true` if either one or both of the operands are equal to `true` but will evaluate to `false` if both operands are equal to false. Finally, the expression (`!`*operand*) will evaluate to `true` if the operand is equal to `false` and vice versa. For example, the expression ((10 > 3) `&&` (10 > 11)) will evaluate to `false`, whereas ((10 > 3) `&&` `!`(10 > 11)) is equal to `true`.

A.3.2. *The* `for` *and* `while` *loops*

The `for` loop allows for a single piece of code to be executed multiple times with slight changes at each iteration. The following simple example sums all the numbers from 1 to 10:

```
int sum = 0;
for(int i=1; i<=10; i++)
{
    sum += i;
}
```

A `for` loop takes the form `for`(*initialisation*; *condition*; *increment*). In typical usage, the initialisation typically sets up the initial value of a loop counter variable (`i` in the above example). The condition is checked at the start of each loop iteration; the loop continues to run until it reaches an iteration where this condition evaluates to `false`. Finally, the increment specifies the amount by which the loop counter is incremented at the end of each iteration.

The `while` loop performs a similar function, executing a code block repeatedly while a condition is `true`; however, the loop control variable must be dealt with manually.

```
int sum = 0, i = 0;
while(i<=10)
{
    sum += i;
    i++;
}
```

A.4. Functions

A function is a sequence of statements that perform a specific task, packaged as a unit. As we have already seen, the function `main()` is the entry point for any C++ program, but a program may also make use of any number of additional functions. Each function can take a pre-specified number of *input parameters*, each of a specified type, carry out a series of instructions using these parameters and optionally return a value of a specified type to the code that called it.

```
int biggest(int number1, int number2)
{
    if(number1 > number2)
    {
        return number1;
    }
    else
    {
        return number2;
    }
}
```

This example function takes two integers as input and returns the value of the larger of the two (also an integer). The types of input parameters do not have to be the same, nor does the return type have to be the same as that of any of the input parameters. If the function does not return a value, the return type is specified as `void`. When the `return` keyword is encountered in a function, a copy of the specified value is returned to the calling code, which also returns control to the calling code; no other statements in the function will be executed. The above function may be called as follows:

```
int big=0, a=15, b=34;
big = biggest(a, b);    // big = 34
```

There are two basic methods of passing an argument to a function. The first as used above is the pass-by-value mechanism, in which a copy of the input parameter is made and it is this copy that is referred to within the body of the function. Under this mechanism, the function cannot alter the value of the input variables (though it can alter the value of the copy), i.e., the values of the variables `a` and `b` could not be altered by the function in the above example. The other method is the pass-by-reference method; the example function can be altered to use this mechanism by changing the prototype (the first line) to

```
int biggest(int& number1, int& number2)
```

The input parameters are now of type reference-to-int. Under this mechanism, the parameters inside the function refer directly to the variables that were passed in and so their values may be directly altered by the function. Further, the values of the input variables do not need to be copied when passing them to the function. This is useful if you are passing, e.g., an array or a vector, as it prevents all the elements in either structure from being copied.

In order to call a function from within `main()`, it must be defined before the definition of `main()` appears in the source code. Alternatively a forward declaration may be used, wherein the function's prototype is listed before the definition of `main()`, and its declaration afterwards (still with the prototype attached):

```
#include <iostream>

using namespace std;

int biggest(int number1, int number2);

int main()
{
    int big=0, a=15, b=34;
    big = biggest(a, b);
    cout << big << " is the biggest";
}

int biggest(int number1, int number2)
{
    if(number1 > number2)
    {
        return number1;
    }
    else
    {
```

```
        return number2;
    }
}
```

References

[1] A. Alexandrescu. *Modern C++ Design: Applied Generic and Design Patterns* (Addison Wesley, Boston, 2001).

[2] N. M. Josuttis. *The C++ Standard Library: A Tutorial and Reference*, 2nd ed. (Addison Wesley, Boston, 2012).

[3] M. Gregoire. *Professional C++*, 4th ed. (Wrox Press, 2018).

[4] S. Meyers. *Effective C++*, 3rd ed. (Addison Wesley, Boston, 2005).

[5] H. Schildt. *C++: A Beginner's Guide*, 3rd ed. (McGraw-Hill Osborne, Berkley, 2014).

[6] B. Stroustrup. *The C++ Programming Language*, 4th ed. (Addison Wesley, Boston, 2013).

[7] H. Sutter. *Exceptional C++* (Addison Wesley, Boston, 1999).

Appendix B: Microdisc Program

The following is a simple C++ program to simulate the cyclic voltamme-try of the one-electron reduction at a disc microelectrode. As with the macrodisc simulation (Subsection 3.4.1), the program begins by specifying the values of the input parameters (θ_i, θ_v, σ, $h0$, ω and $\Delta\theta$) and then calculating the values of some other parameters (ΔT, $T_{\max}$, $R_{\max}$, $Z_{\max}$ and the number of timesteps). Next we create the expanding spatial grids for both the Z- and R-directions and use these to calculate the total num-ber of spacesteps in each direction (m and n respectively) and then create two two-dimensional arrays in which to store (1) the concentrations at the previous timestep and (2) the concentrations at the current timestep/$\delta_{i,j}$ values.

The next step is to calculate the values of the α_i, β_i and γ_i coefficients for both the R- and Z-directions and then those of the modified γ_i values. Note that because there are two surface boundary conditions for the Z-sweep (electroactive disc surface and electroinactive supporting surface), there are necessarily two sets of modified γ_i coefficients.

The main simulation loop proceeds as follows: at the start of each timestep, the new value of the potential, θ, is calculated and the solution (concentration grid) from the previous timestep is copied into the array C_. Next is the Z-implicit sweep, each column of which is solved using the Thomas algorithm. After this, the concentration grid is again copied so that it can be used in the next half-timestep, and the current is recorded and output to a text file. The current should specifically be recorded af-ter the Z-sweep because the implicit diffusion is calculated perpendicular to the electrode in that sweep. The R-sweep follows in which each row is solved using the Thomas algorithm.

The program uses OpenMP in order to implement multithreading [1, 2]. Assuming the use of a compatible compiler,[1] the use of OpenMP can vastly

---

[1] See https://www.openmp.org/resources/openmp-compilers-tools/ for a list of compat-ible compilers.

decrease the running time of the simulation (in proportion to the number of available processor cores). To use it, the header `<omp.h>` must be included and the function `omp_set_num_threads()` called to specify the number of threads that it should attempt to use. While OpenMP allows for more advanced usage, for our relatively modest needs, we simply prefix the `for` loop that we wish to parallelise with the directive `#pragma omp parallel for`, as shown in the program below. Multiple loop iterations will then run concurrently, the number of simultaneous iterations being equal to the number of threads.

As the order of execution of the loop iterations is now essentially non-deterministic (in that we cannot say ahead of time which iteration will execute when), it is very important that the data accessed in each loop iteration is orthogonal, i.e., that the results of one iteration do not depend in any way on the results of any other previous iteration.

```cpp
#include <fstream>
#include <vector>
#include <cmath>
#include <omp.h>

int main(int argc, char* argv[])
{
    // Set number of threads to be used by OpenMP
    omp_set_num_threads(4);

    // Specify simulation parameters
    double theta_i = 20.0;
    double theta_v = -20.0;
    double sigma = 1000.0;

    double h0 = 1e-4;
    double omega = 1.08;
    double deltaTheta = 0.01;

    // Determine other parameters
    double deltaT = deltaTheta / sigma;
    double maxT = 2.0 * fabs(theta_v - theta_i) / sigma;
    double maxR = 6.0 * sqrt(maxT) + 1.0;
    double maxZ = 6.0 * sqrt(maxT);
```

```cpp
int t = (int)( maxT / deltaT );   // number of timesteps

// Make Z grid
std::vector<double> Z;
double h = h0;
Z.push_back(0.0);
while( Z.back() <= maxZ ) {
    Z.push_back( Z.back() + h );
    h *= omega;
}
int m = Z.size(); // number of spacesteps (Z)

// Make R grid
std::vector<double> R;
h = h0;
R.push_back(0);
while( R.back() < 0.5 ) {
    R.push_back( R.back() + h );
    h *= omega;
}
R.back() = 0.5;

for(int i = R.size()-2; i>=0; i--) {
    R.push_back( 1 - R[i] );
}
int n_e = R.size(); // number spacesteps over electrode

h = h0;
while( R.back() <= maxR ) {
    R.push_back( R.back() + h );
    h *= omega;
}
int n = R.size(); // number of spacesteps (R)

// Make concentration grids
double** Ck = new double*[n];
double** C_ = new double*[n];

for(int i=0; i<n; ++i)
```

```
    {
        Ck[i] = new double[m];
        C_[i] = new double[m];

        // set initial concentration
        for(int j=0; j<m; j++) {
            Ck[i][j] = 1.0;
            C_[i][j] = 1.0;
        }
    }


    // Create and set Thomas coefficients
    std::vector<double> z_al(m,0.0), z_be(m,0.0), z_ga(m,0.0);
    std::vector<double> r_al(n,0.0), r_be(n,0.0), r_ga(n,0.0);
    std::vector<double> ga_modZ1(m, 0.0), ga_modZ2(m, 0.0);
    std::vector<double> ga_modR(n, 0.0);

    for(int j=1; j<m-1; j++)
    {
        z_al[j] = 2.0 / ( (Z[j+1]-Z[j-1]) * (Z[j] - Z[j-1]) );
        z_ga[j] = 2.0 / ( (Z[j+1]-Z[j-1]) * (Z[j+1] - Z[j]) );
        z_be[j] = -z_al[j] - z_ga[j] - 2.0 / deltaT;
    }

    for(int i=1; i<n-1; i++)
    {
        r_al[i] = ( 1.0 / (R[i+1]-R[i-1]) )
            * ( 2.0/(R[i] - R[i-1]) - 1.0/R[i] );
        r_ga[i] = ( 1.0 / (R[i+1]-R[i-1]) )
            * ( 2.0/(R[i+1] - R[i]) + 1.0/R[i] );
        r_be[i] = -r_al[i] - r_ga[i] - 2.0 / deltaT;
    }

    // Modify gamma coefficients for Z sweep
    ga_modZ1[0] = 0.0; // electrode boundary condition
    for(int j=1; j<m-1; j++) {
        ga_modZ1[j] = z_ga[j]
            / (z_be[j] - ga_modZ1[j-1] * z_al[j]);
    }
```

```
ga_modZ2[0] = -1.0; // no-flux boundary condition
for(int j=1; j<m-1; j++) {
    ga_modZ2[j] = z_ga[j]
        / (z_be[j] - ga_modZ2[j-1] * z_al[j]);
}


// Modify gamma coefficients for R sweep
ga_modR[0] = -1.0; // boundary condition
for(int i=1; i<n-1; i++) {
    ga_modR[i] = r_ga[i]
        / (r_be[i] - ga_modR[i-1] * r_al[i]);
}


// Open file to output CV
std::ofstream CV("CV_Output.txt");

// BEGIN SIMULATION
double Theta = theta_i;
for(int k=0; k<t; k++)
{
    if( k < t/2 ) { Theta -= deltaTheta; }
    else          { Theta += deltaTheta; }

    //copy concentration grid
    for(int i=0; i<n; i++) {
        for(int j=0; j<m; j++) {
            C_[i][j] = Ck[i][j];
        }
    }


    //--- Z SWEEP---
    #pragma omp parallel for
    for(int i=1; i<n-1; i++)
    {
        Ck[i][m-1] = 1.0;
        for(int j=1; j<m-1; j++)
        {
            Ck[i][j] = - C_[i-1][j] * r_al[i]
                - C_[i][j] * (-r_al[i] - r_ga[i])
```

```
            - C_[i][j] * 2.0/deltaT
            - C_[i+1][j] * r_ga[i];
    }

    // Set surface deltas and pointer to gamma_mod
    std::vector<double>* ga_modZ;
    if(i < n_e) {
        Ck[i][0] = 1.0 / (1.0 + exp(-Theta));
        ga_modZ = &ga_modZ1;
    }
    else {
        Ck[i][0] = 0.0;
        ga_modZ = &ga_modZ2;
    }

    // Modify deltas
    for(int j=1; j<m-1; j++)
    {
        Ck[i][j] = ( Ck[i][j] - Ck[i][j-1] * z_al[j] )
            / ( z_be[j] - (*ga_modZ)[j-1] * z_al[j] );
    }

    //solve by back substitution
    Ck[i][m-1] = 1.0;
    for(int j=m-2; j>=0; j--) {
        Ck[i][j] = Ck[i][j]-(*ga_modZ)[j]*Ck[i][j+1];
    }
}

//copy concentration grid
for(int i=0; i<n; i++) {
    for(int j=0; j<m; j++) {
        C_[i][j] = Ck[i][j];
    }
}

//Output current
double flux = 0.0;
for(int i=1; i<n_e; i++) {
```

```
        double J2 = (Ck[i][1] - Ck[i][0]) * R[i];
        double J1 = (Ck[i-1][1] - Ck[i-1][0])*R[i-1];
        flux -= (0.5/h0)*(J2+J1)*(R[i] - R[i-1]);
    }
    CV << Theta << "\t" << flux << "\n";

    //--- R SWEEP---
    #pragma omp parallel for
    for(int j=1; j<m-1; j++)
    {
        // set deltas
        Ck[0][j] = 0;
        Ck[n-1][j] = 1.0;
        for(int i=1; i<n-1; i++)
        {
            Ck[i][j] = - C_[i][j-1] * z_al[j]
                - C_[i][j] * (-z_al[j] - z_ga[j])
                - C_[i][j] * 2.0/deltaT
                - C_[i][j+1] * z_ga[j];
        }

        // modify deltas
        for(int i=1; i<n-1; i++)
        {
            Ck[i][j] = ( Ck[i][j] - Ck[i-1][j] * r_al[i] )
                / ( r_be[i] - ga_modR[i-1] * r_al[i] );
        }

        //solve by back substitution
        for(int i=n-2; i>=0; i--) {
            Ck[i][j] = Ck[i][j] - ga_modR[i]*Ck[i+1][j];
        }
    }
  }
}
```

References

[1] L. Rohit Chandra, D. Dagum, D. Maydan, J. Kohr, and R. M. McDonald. *Parallel Programming in OpenMP* (Morgan Kaufmann, San Francisco, 2000).
[2] B. Chapman, G. Jost, and R. van der Pas. *Using OpenMP: Portable Shared Memory Parallel Programming* (The MIT Press, Cambridge, 2007).

Index